98-REM-1

A Dynamic Model to Assess Microbial Health Risks Associated with Beneficial Uses of Biosolids — Phase 1

by:

John M. Colford, Jr., M.D, Ph.D.
School of Public Health, University of California, Berkeley

Don M. Eisenberg, Ph.D.
Eisenberg, Olivieri and Associates, Inc

Joseph N.S. Eisenberg, Ph.D.
School of Public Health, University of California, Berkeley

James Scott, MPH
School of Public Health, University of California, Berkeley

Jeffrey A. Soller
Eisenberg, Olivieri and Associates, Inc

2003

Water Environment Research Foundation

The Water Environment Research Foundation, a not-for profit organization, funds and manages water quality research for its subscribers through a diverse public-private partnership between municipal utilities, corporations, academia, industry, and the federal government. WERF subscribers include municipal and regional water and wastewater utilities, industrial corporations, environmental engineering firms, and others that share a commitment to cost-effective water quality solutions. WERF is dedicated to advancing science and technology addressing water quality issues as they impact water resources, the atmosphere, the lands, and quality of life.

For more information, contact:
Water Environment Research Foundation
601 Wythe Street
Alexandria, VA 22314-1994
Tel: (703) 684-2470
Fax: (703) 299-0742
www.werf.org
werf@werf.org

This report was co-published by the following organizations:
For non-Subscriber sales information, contact:

Water Environment Federation
601 Wythe Street
Alexandria, VA 22314-1994
Tel: (800) 666-0206
Tel: (703) 684-2452
Fax: (703) 684-2492
www.wef.org
pubs@wef.org

IWA Publishing
Alliance House, 12 Caxton Street
London SW1H 0QS, United Kingdom
Tel: +44 (0) 20 7654 5500
Fax: +44 (0) 20 7654 5555
www.iwapublishing.com
publications@iwap.co.uk

Library of Congress Catalog Card Number: 2003100317
Printed by Ashford Overload, Southampton, UK
WERF ISBN: 1-893664-73-2 IWAP ISBN: 1-84339-634-3

University of California, Berkeley

The research on which this report is based was funded, in part, by the United States Environmental Protection Agency through Cooperative Agreement No. CR-825237 with the Water Environment Research Foundation (WERF). Unless an EPA logo appears on the cover, this report is a publication of WERF, not EPA. Funds awarded under the Cooperative Agreement cited above were not used for editorial services, reproduction, printing, or distribution.

ACKNOWLEDGMENTS

The authors thank the members of the project's Health Advisory Committee: Phillip Berger, Robert Cooper, Walter Jakubowski, Mark Sobsey, and George Tchobanoglous.

Report Preparation

Principal Investigator:

John M. Colford, Jr., M.D., Ph.D.
School of Public Health, University of California, Berkeley

Project Team:

Joseph N.S. Eisenberg, Ph.D.
School of Public Health, University of California, Berkeley

James Scott, MPH
School of Public Health, University of California, Berkeley

Jeffrey A. Soller
Eisenberg, Olivieri and Associates, Inc

Don M. Eisenberg, Ph.D.
Eisenberg, Olivieri and Associates, Inc

Project Subcommittee

Cecil Lue-Hing, D.Sc., PE, DEE, Chair
Cecil Lue-Hing & Associates, Inc.

Sydney Munger, Research Council Liaison, Co-Chair
Parametrix, Inc.

Richard Danielson, Ph.D.
Biovir

Salvador Sedita, Ph.D.

James E. Smith, Jr. D.Sc.
USEPA

William Yanko

Water Environment Research Foundation Staff

Research Program Director: Jami Montgomery

ABSTRACT AND BENEFITS

Abstract:

Maximum allowable levels for *chemical* contaminants in biosolids were developed for the Part 503 rule using risk-based methodologies. However, maximum allowable levels of *microbiological* contaminants in the Part 503 rule were based on specific treatment methodologies rather than risk levels, because it was determined at that time that risk assessment methodologies were not sufficiently developed. Given the current interest in the beneficial uses of biosolids and the projected rapid growth of biosolids reuse, there is increasing interest in the development of a microbial risk assessment methodology for regulatory and operational decision making.

This document presents a methodology for assessing risks to human health from pathogens via exposure to biosolids. The methodology integrates two fundamental components: an exposure assessment component and a health risk component. The exposure assessment component is used to quantify pathogen levels in the environment and serves as input to the health effects component. The health effects component is used to quantify health risks using a model that explicitly accounts for properties unique to an infectious disease process, specifically secondary transmission and immunity.

To demonstrate the applicability of these risk-based methods developed for biosolids exposure, numerical simulations were carried out for a case study example in which the route of exposure was direct consumption of biosolids-amended soil. The output from the case study yielded a decision tree that differentiates between conditions in which the risk from biosolids exposure is high and those conditions in which the relative risk from biosolids is low. This decision tree illustrates the interaction among the important factors in quantifying risk. For the case study example, those factors include biosolids treatment processes, the pathogen shedding rate of infectious individuals, secondary transmission and immunity. Further work in determining biosolids exposures is required before this methodology can be used in a comprehensive risk assessment.

Benefits :

- This project provides a structure for estimating risks that is consistent with the current understanding of infectious disease processes and a methodology for estimating the number of cases in a population that may be associated with biosolids exposure. This information can be used to develop risk-based regulations for biosolids-related exposures.
- The microbial risk assessment methodology provides estimates on the uncertainty in risk that are useful for decision making.
- The model is flexible and allows for the incorporation of future data related to biosolids, pathogens, or both.

Key Words: risk assessment, pathogens, land application, biosolids

TABLE OF CONTENTS

LIST OF TABLES

LIST OF FIGURES

LIST OF ACRONYMS

AR	attributable risk
CART	classification and regression tree
CI	cumulative incidence
EPA	United States Environmental Protection Agency
HAC	health advisory committee
HEI	highly exposed individual
ILSI	International Life Sciences Institute
IU	infectious units
MPN	most probable number
NRC	National Research Council
PFU	plaque-forming units
PSC	Project Subcommittee
RT-PCR	reverse transcriptase polymerase chain reaction
TS	total solids
WERF	Water Environment Research Foundation

EXECUTIVE SUMMARY

This document presents a methodology for assessing the risk to human health from exposure to pathogens found in biosolids. This methodology is consistent with the epidemiology of infectious disease transmission and the general approach recommended by the U.S. Environmental Protection Agency (ILSI 1996) for carrying out microbial risk assessment. The methodology takes a three-step approach to assess risks to human health from exposure to pathogens:

- *Problem formulation:* This step includes the development of a conceptual model that represents both the exposure and health effects components of a risk assessment.
- *Data collection:* Based on the conceptual model, the necessary exposure and health effects data are collected.
- *Risk characterization:* Based on the conceptual model and the collected data, a disease transmission model is used to estimate risk. This characterization step contains a sensitivity analysis for proper interpretation of the risk estimates.

ES.1 The 503 Rule

In 1993, the U.S. Environmental Protection Agency (EPA) published in the *Federal Register* the Part 503 rule (USEPA 1993b). The Part 503 rule was a "comprehensive, risk-based rule to protect public health and the environment from reasonably anticipated adverse effects of pollutants that may be present in biosolids that are used or disposed" (USEPA 1995). Pollutant limits for chemical constituents in biosolids were based on the results of risk assessments that were conducted to identify what, if any, risks were associated with the use or disposal of biosolids. Pollutant limits for pathogens, however, were not based on risk assessments; rather they were based on either performance, technology-based standards, or on management and record-keeping practices shown in the past to protect human health and the environment. The regulation of pathogens in the final Part 503 rule was not based on risk assessment "because methodologies had not been developed sufficiently to make such calculations" (USEPA 1995). The value of a health-risk-based approach for both regulatory and management practices is that decision-making can include consideration of the risk of disease associated with different control measures. Evaluation of operational or technology-based approaches can also consider health risk implications.

During the past decade there have been a number of advancements in microbial risk assessment including: the development of a comprehensive approach using the chemical risk paradigm (Regli et al. 1991); the compilation of information and data as well as an overview of the chemical risk paradigm approach (Haas et al. 1999); the development by EPA of a microbial risk framework (ILSI 1996); and an extension of this EPA framework to include a population-level, disease-transmission-based framework (Chick et al. 2001; Eisenberg et al. 1996; Eisenberg et al. 1998; EOA 1999).

Chemical risk assessment has been used extensively since the National Research Council (NRC) report on risk assessment (1983). By comparison, microbial risk assessment is in its infancy. Because of the earlier development of chemical risk assessment, attempts to provide a quantitative assessment of human health risks associated with the ingestion of pathogens have generally focused on static models that calculate the probability of individual infection or disease as a result of a single exposure event (Haas 1983; Regli et al. 1991). The approach generally employed for these assessments has been based upon the NRC chemical risk model and, as such, does not address a number of properties that are unique to infectious disease transmission, such as person-to-person disease transmission and immunity.

EPA realized that infectious disease processes had features significantly different from disease processes caused by chemical exposures, and began preliminary work by sponsoring a workshop in 1996, from which a working group developed a conceptual approach for microbial risk assessment (ILSI 1996). In contrast to the previous approach, EPA microbial risk approach allows for the incorporation of pathogen-specific epidemiological data, such as incubation period, immune status, duration of disease, and the rate of symptomatic development. However, EPA's approach falls short of explicitly incorporating features of the disease transmission process that would be necessary to highlight the distinction between infectious disease risk and a more conventional static risk process (EOA 1999) such as those described immediately below.

Infectious disease transmission processes can be distinguished from chemical-agent disease processes in two main ways. First, pathogens are transmitted from host to host either directly (through close contact) or indirectly (through environmental sources). The principal feature of this characteristic is that all pathways of transmission are integrated and therefore cannot be considered independent processes. For example, a child may be infected due to exposure from biosolids and transmit the infection through person-to-person contact in household, school, or day care settings. In this example, the person-to-person transmission was dependent on and was a result of biosolids exposure. Second, after initial exposure to a pathogen, a host may exhibit a certain degree of protection from disease (immunity). The degree to which a pathogen is circulating within a population and the extent to which the population has previously been exposed to the pathogen determines the degree to which protection is conferred on the population.

This investigation involved the development and demonstration of a risk-based methodology that explicitly encompasses the unique features of an infectious disease process, as well as the inclusion of biosolids as one of the potential routes of exposure to pathogens.

ES.2 Methodology Development

The methodology integrates two fundamental components: an exposure assessment component and a health effects component. The exposure assessment component is used to quantify the pathogen levels in the environment. Based on the exposure assessment component, the health effects component is used to quantify the health risks using a disease transmission model.

The risk estimates provided by the disease transmission model are manifested at the population level. This therefore dictates a population-level measure of disease risk such as cumulative incidence, defined as the number of new cases of illness observed over a defined period of time divided by the number of persons initially at risk (Hennekens et al. 1987). To estimate the risk specifically associated with exposure to biosolids requires the estimate of the attributable risk. Attributable risk is defined as the cumulative incidence with biosolids exposure removed, subtracted from the cumulative incidence with biosolids exposure present, keeping all other factors constant (Hennekens et al. 1987). Describing risk in terms of attributable risk is useful when there are competing pathways of infection since this measure specifically quantifies to what extent a particular exposure (in this case biosolids) are causally related to a disease outcome. The cumulative incidence presents the overall number of cases occurring in the population, whereas the attributable risk presents the number of cases associated with a particular exposure.

The results obtained from a chemical risk assessment approach are also measures of the number of cases associated with a particular exposure. The chemical risk approach defines risk as the probability of contracting disease per-person for a given time period. This is a cumulative risk estimate. Since the assumption is that secondary transmission is negligible and that any other disease transmission pathway is independent of the particular pathway being studied, this cumulative risk estimate is the same as the attributable risk.

A major assumption behind this project was that it was important to account for factors such as secondary transmission and the interdependencies of pathways. One method to incorporate these is to use a disease transmission model. The disease transmission model can provide a risk estimate with biosolids exposure present and a risk estimate with biosolids exposure not present. It is the difference of these two risk estimates that is then the attributable risk associated with biosolids exposure.

The parameters used in the model represent factors involved in the exposure and disease processes. The model structure defines these factors and their relationship to risk estimation. The values used to identify these parameters were obtained from the literature and are described in detail in Chapter 2.0. Chapter 3.0 summarizes this information and presents simulations demonstrating how risk may be assessed employing a disease transmission model.

ES.3 Demonstration of Methodology

To demonstrate the utility of the methodology, a risk characterization, as well as a sensitivity analysis of the risk estimates, was conducted. Risk characterizations are conducted for specific environmental conditions. The methodology is demonstrated using a hypothetical but realistic scenario. This scenario is not meant to account for all types of biosolids exposures. Rather, it is simply meant to provide a baseline construct from which to illustrate the utility of the methodology.

Briefly, the hypothetical community in the case study was assumed to be closed to external sources of pathogens; i.e., pathogens in the environment were generated by infectious community members shedding pathogens. For simplicity, the community members were assumed to be homogeneous with respect to exposure and susceptibility to disease, although the model can also incorporate a heterogeneous population. The assessment was carried out using enteroviruses as the example pathogen. Exposure to enteroviruses was assumed to occur via the ingestion of biosolids-amended soil, and the enterovirus pathogen was assumed to be homogeneous; i.e., distinct characteristics of the different enterovirus strains were not modeled. With respect to the environmental conditions, the pathogens transported to the biosolids exposure site were attenuated by a wastewater treatment facility, natural die-off, dilution, and biosolids processing.

ES.4 Risk Characterization/Sensitivity Analysis

Model simulation output was used to map out the conditions under which the attributable risk to biosolids exposure was high as well as those conditions for which it was low. The simulation results indicated that there were four important factors that most strongly affected the level of risk:

1. The ratio of non-biosolids environmental exposure to biosolids environmental exposure (X_e/X_b);
2. The rate of infectious individuals shedding pathogens into the environment (ϕ). This pathway of transmission can be controlled through both wastewater and biosolids treatment;
3. The rate of person-to-person (secondary) transmission (β). Increased secondary transmission resulted in a decrease in attributable risk due to biosolids (e.g., a highly infectious pathogen will manage to infect an individual through another pathway if infection is prevented from the biosolids pathway); and
4. The duration of immunity (γ). An increase in duration of immunity also resulted in a decrease in attributable risk due to biosolids.

The sensitivity analysis suggested that attributable risk is low regardless of other parameter values when exposure from biosolids is small compared to exposure from other environmental sources, if the pathogen-shedding rate is relatively high or low (not medium), and if the secondary transmission rate is relatively high.

ES.5 Conclusions

This project developed and demonstrated a risk-based methodology for assessing the risk to human health from exposure to pathogens via biosolids. The simulation study produced results that highlight some of the distinctions between using a chemical risk model and a disease transmission model. Although the methodology is appropriate for assessing risks to human health for specified routes of exposure, a substantial amount of research may be necessary before this methodology will be useful for characterizing the actual risk from exposure to pathogens from biosolids. Comprehensive risk assessments will require extensive exposure determinations though either primary data collection or exposure modeling.

The methodology developed in this project:

- Provides an initial screening of a given scenario, identifying broad conditions for high and low attributable risk situations;
- Supplies information on the impact of implementing specific control measures, e.g., estimates can be provided for the change in risk associated with a 50% increase in storage time;
- Identifies estimates for which more data are needed; e.g., the uncertainty in specific pathogen die-off estimates may be responsible for much of the uncertainty in risk estimates. The identification of these data needs can then help to set future research needs.

Future work should now be focused on using this methodology with more realistic and representative scenarios and on identifying crucial gaps in existing data.

CHAPTER 1.0

INTRODUCTION

Biosolids are the finished solid products that are generated via the treatment of wastewater and subsequent treatment of sludge. Recently there has been increasing interest in the beneficial uses of biosolids, such as large-scale application to agricultural land as well as application at a smaller scale for use in home gardens. This was motivated by the potential for both environmental (Tenenbaum 1997) and economic (Faust and Oberst 1996, Riggle 1996) gains. This rapid growth in the public use of biosolids has raised concerns about the abilities of the facilities that handle them to meet control and monitoring standards (Hay 1996, Logan 1995). Pollutants that may be present in biosolids are regulated by the United States Environmental Protection Agency (EPA), under what is commonly known as the Part 503 rule. The sections of the rule that addressed chemicals were based on the results of risk assessments. The sections of the rule that addressed pathogens were not based on risk assessments; rather they were based on performance or technology-based standards, management, and record keeping practices. The first step in addressing concerns related to microbial pathogens and biosolids is to provide a risk-based standard rather than a technology- or treatment-based standard.

This investigation was carried out in response to a Water Environment Research Foundation (WERF) request for proposals "to develop a scientifically-based, practical methodology to assess the relative risk of pathogenic microorganisms to human health through exposure to biosolids used for beneficial purposes." The goal of this investigation was to provide a formal structure for assessing risks of microbial pathogens that may be used in a manner similar to the current planning and regulatory uses of the commonly accepted chemical risk assessment methodology. Characterizing the risk to human health associated with exposure to pathogens within biosolids involves three steps:

- *Problem formulation:* This includes the development of a conceptual model that represents both the exposure and health effects components of a risk assessment.
- *Data collection:* Based on the conceptual model the necessary exposure and health effects data are collected.
- *Risk characterization:* Based on the conceptual model and the collected data, a disease transmission model is used to estimate risk. This characterization step contains sensitivity analyses required for proper interpretation of the risk estimates.

The following section provides a brief background on 503 rule development, summarizing the basis for the existing regulation of both chemicals and pathogens in biosolids. As mentioned previously, although the regulations for chemical exposures were based on a risk assessment, the regulations for pathogens were not. The last section of this chapter describes a risk assessment conducted by EPA for land application of biosolids but never used in the 503 regulations. The methodology used in this risk assessment was based on an approach developed for assessing the risk of exposure to chemical contaminants in the early 1980s. This last section ends with a summary of the emergence of an alternative model for assessing risk from exposure to microbial pathogens.

1.1 The Part 503 Rule: Basis for Biosolids Regulations

This section covers EPA's previous work related to characterizing the risk associated with contaminants (both chemical and microbial) in biosolids, describes the basis for the existing regulation of pathogens in biosolids, and summarizes the existing regulations pertaining to pathogens in biosolids.

EPA developed a risk-based rule to protect public health and the environment from reasonably anticipated adverse effects of pollutants that may be present in biosolids that are used or disposed. The rule was published in the *Federal Register* in 40 CFR Part 503 and is commonly known as the Part 503 rule. Pollutant limits for chemical constituents in biosolids were developed through risk assessment, whereas pollutant limits for pathogens were performance or technology based.

Following is a brief summary of the risk assessment work that led to the Part 503 rule for chemicals and the preliminary pathogen risk assessments conducted for biosolids applied to land (USEPA 1989c; USEPA 1991a; USEPA 1991b; USEPA 1992). Although preliminary pathogen risk assessments were carried out prior to the adoption of the Part 503 rule and are described in this section, it should be noted that those assessments were not used directly in the Part 503 rule regulations (USEPA 1995).

1.1.1 The Regulation of Chemical Contaminants in Biosolids

Extensive risk assessments were conducted for land application, surface disposal, and incineration of biosolids in the development of the Part 503 rule. The general procedure used was based on the NRC chemical risk paradigm. The basic approach used in the biosolids risk assessments was to identify highly exposed individuals (HEI) for different pollutants of concern through specific exposure pathways (USEPA 1995). The assumptions used to estimate exposure to an HEI involved combinations of "high-end" and "mid-range" assumptions in an attempt to provide realistic exposure estimates and, at the same time, conservative protection.

Initially, EPA developed a list of approximately 200 pollutants as possible candidates for the risk assessments. Based on the recommendations of an expert panel, that list was reduced to 50 pollutants. Using the hazard index method, defined as

$$\frac{\text{concentration in soil, plant or animal tissue, water, or air}}{\text{lowest concentration known to be toxic to the organism being evaluated}}$$

environmental profiles were used to evaluate the impact of each of the 50 pollutants by comparing its concentration in the environment with established human health or regulatory criteria. The pathways investigated were land application, surface disposal, and incineration. Pollutant/pathway combinations with hazard indices less than 1 were dropped from further consideration. Each pollutant/pathway combination with a hazard index greater than 1 was further evaluated by excluding background exposure to the pollutant, leaving the impacts from the pollutant in biosolids only. This process resulted in reducing the list of pollutants to 22 for assessing risks due to land application of biosolids, 16 for surface disposal, and 14 for incineration.

The biosolids risk assessments conducted for the final Part 503 rule were conducted as follows:

- Cancer risks for all biosolids use or disposal practices were regulated at 10^{-4};
- Formal pathway exposure assessments were conducted for land application and surface disposal practices;
- The most stringent of the pollutant limits for each pollutant from all pathways of exposure was identified as the "limiting pathway," and this value was used as the pollutant limit;
- Organic pollutants for land application and surface disposal were not regulated because they occurred in biosolids in the United States at low levels and, therefore, were assumed not to pose significant risks to public health or the environment; and
- Cadmium, copper, lead, mercury, and chromium were not regulated in surface-disposed biosolids because none were expected to exceed the levels identified in the risk assessment.

Fourteen exposure pathways were evaluated for land application of biosolids. Pollutant limits were developed using the limiting pathway for both inorganic and organic contaminants. The limiting pathways for the pollutants identified included: a child eating biosolids, plant phytotoxicity, an animal eating feed, an adult eating animal products, an adult eating fish, and an adult drinking surface water.

1.1.2 The Regulation of Pathogens in Biosolids

The regulation of pathogens in the final Part 503 rule is not based on risk assessment "because methodologies had not been developed sufficiently to make such calculations" (USEPA 1995). Instead, the Part 503 pathogen operational standard included pathogen controls and monitoring requirements for all biosolids; and crop harvesting, animal grazing, and site access restrictions for certain biosolids. This operational standard, which requires that the pathogens in biosolids be reduced to below detectable levels, was based on "extensive experimental data and years of experience and, in the judgment of EPA, is protective of public health and the environment" (USEPA 1995).

The Part 503 regulation contains two classes of pathogen reduction: Class A and Class B. Class A pathogen reduction alternatives render the biosolids "virtually pathogen free after the treatment" (USEPA 1994). Class B pathogen reduction alternatives "significantly reduce but do not eliminate the pathogens" (USEPA 1994). Land appliers who apply biosolids certified as Class A have no requirements relative to pathogens. If the biosolids are Class B, site restrictions must be imposed to allow time for natural processes to further reduce pathogen levels. There are a number of alternative methods that may be used to determine whether biosolids may be classified as Class A or Class B. Interested readers are referred to "A Plain English Guide to the EPA Part 503 Biosolids Rule" (USEPA 1993a).

1.2 Microbial Risk Assessment

1.2.1 Pathogen Risk Assessment for Land Application of Biosolids

Although not used directly in the development of the Part 503 regulations, EPA did develop, between the mid-1980s and the early 1990s, preliminary assessments of risk to human health posed by parasites, bacteria, and viruses in municipal biosolids applied to land as fertilizer or soil conditioner (USEPA 1991a; USEPA 1991b; USEPA 1992). The motivation for these risk assessments came from section 405 of the Clean Water Act, which required EPA to develop and issue regulations that identified uses of sludge and biosolids including disposal, specific factors to be taken into account in determining the measures and practices applicable to each use or disposal, and concentrations of pollutants that interfere with each use or disposal. These risk assessments provide a starting point for the work presented in this document by detailing relevant applications and exposure pathways, identifying the most critical data gaps, and recommending research priorities.

The five municipal management practices included in the preliminary risk assessment were application of liquid-treated sludge to commercial crops for human consumption, grazed pastures, and crops to be processed before being consumed by animals. Moreover, application of dried or composted biosolids to residential vegetable gardens and residential lawns were also included in the preliminary risk assessments.

Exposure pathways (migration routes from or within the application site to a receptor) for biosolids applied to land included the following:

- Inhalation and ingestion of aerosols for the application of liquid sludge or wastewater;
- Inhalation and ingestion of windblown or mechanically generated particles;
- Swimming in a pond fed by surface water runoff;
- Direct contact with biosolids-contaminated soil or crops;
- Drinking water from an offsite well;
- Consumption of vegetables grown in biosolids-amended soil; and
- Consumption of meat or milk from cattle grazing on or consuming forage from biosolids-amended fields.

Results from the preliminary risk assessments confirmed literature observations that pathogens in treated biosolids could present a potential health risk. Specifically, the study concluded that exposure through contaminated surface water, represented by swimming in a pond fed by surface water runoff was the most significant source of exposure. These predicted risks peaked after each rainfall when additional contaminated surface water and soil were washed into the pond.

The following needs were identified to yield more reliable risk assessments for pathogens present in biosolids:

- Simple and accurate standardized methods for detecting and quantifying pathogens in biosolids;
- Improved understanding of low-dose effects and effects on sensitive subjects;
- More accurate persistence, survival, and transport data on all pathogenic agents of concern in biosolids, and an enhanced understanding of the relationship between those rates and the corresponding environmental conditions, sludge treatment, method of biosolids application and the effects of crop cover;
- Additional information on regrowth of bacteria in compost including factors enhancing or limiting growth;
- Development of an index of soil types that would correlate capacity for solute transport and suitability for biosolids application;
- Research on subsurface injection of biosolids and the relative probability of virus transport in groundwater; and
- Epidemiologic studies evaluating enteric viral transmission and whether there is a correlation between bacterial infections and bacterial aerosols.

1.2.2 A Risk Assessment Approach for Chemical Exposure

Risk assessment involves the use of factual data to define the potential health effects of exposure to hazardous materials and situations for individuals or populations. Microbial risk assessment involves evaluating the likelihood that an adverse health effect may result from human exposure to one or more potential pathogens (Haas et al. 1999).

The four major steps that define these risk assessments are (NRC 1983):

1. *Hazard identification:* The determination of whether a particular agent is or is not causally linked to particular health effects;
2. *Dose-response identification*: The determination of the relation between the magnitude of exposure and the probability of occurrence of the health effects in question;
3. *Exposure assessment:* The determination of the extent of human exposure before or after application of regulatory controls; and
4. *Risk characterization:* The description of the nature and often the magnitude of human risk, including attendant uncertainty.

1.2.3 A Risk Assessment Approach for Microbial Exposure

A drawback to the chemical risk assessment paradigm with respect to infectious disease transmission is that there are issues unique to infectious disease transmission that the chemical risk paradigm does not incorporate, such as secondary (person-to-person) spread of disease and immunity (Eisenberg et al. 1996, EOA 1995). The limitations of treating infectious disease transmission as a static disease process, with no interaction between those infected or diseased and those at risk, has been illustrated in studies of *Giardia* (Eisenberg et al. 1996), dengue (Koopman and Longini 1994), and sexually transmitted diseases (Koopman et al. 1991). Moreover, from a public health perspective, the probable number of people infected in an exposed population provides more insight and is more meaningful than the probability of individual infection. To address this public health perspective, the probability of individual infection has sometimes been multiplied by the number of exposed individuals in an attempt to predict the disease burden in the population (Anderson et al. 1998). Such an approach, however, may not lead to accurate risk forecasts for the population if immunity from infection and/or person to person transmission are important factors for the pathogen of interest (Brookhart et al. in press; Chick et al. 2001; McDonald et al. 2001).

In 1996, the EPA Office of Water contracted the International Life Sciences Institute to convene a panel of experts to develop a conceptual framework to assess the risks of human disease associated with exposure to waterborne pathogenic microorganisms. The panel acknowledged drawbacks to the chemical risk assessment approach when applied to microorganisms, and developed a generalized approach applicable for pathogens (ILSI 1996). The EPA approach comprises three fundamental phases: problem formulation, analysis, and risk characterization. An interpretation of the EPA approach is shown schematically in Figure 1-1 (EOA 1999).

During the problem formulation phase, a clear statement of the problem was formalized, including the goals and scope of the risk assessment to identify the purpose of the risk assessment, the critical issues to be addressed, and how the results might be used to protect public health. Descriptions of the exposure routes and potential health effects were then developed. Exposure routes included the potential sources and vehicles of transmission. Potential health effects include all epidemiologically relevant information. Using this information, initial pathogen and host characterizations were developed, a literature search was conducted, and a conceptual transmission model was developed.

The conceptual model is an explicit diagram of the causal relationships among the different epidemiological states and exposure to the pathogen of interest. It is used as a starting point for the analysis phase of the risk assessment and later as an interactive tool along with components developed in the analysis phase to initiate the risk characterization.

The analysis phase is an evaluation of the available data and is conducted in parallel with the structure of the conceptual model. The information and data obtained during the problem formulation phase are reviewed and organized into categories describing exposure and host-pathogen interactions. The analysis phase results in exposure and host/pathogen "profiles" which are succinct summaries of the most relevant information and data for the risk assessment, including uncertainty and variability, where possible and appropriate.

In the risk characterization phase of the risk assessment, information from the analysis phase is integrated systematically into the revised models for exposure and health effects. This process results in an estimation of the likelihood of adverse health effects based on the exposure under study.

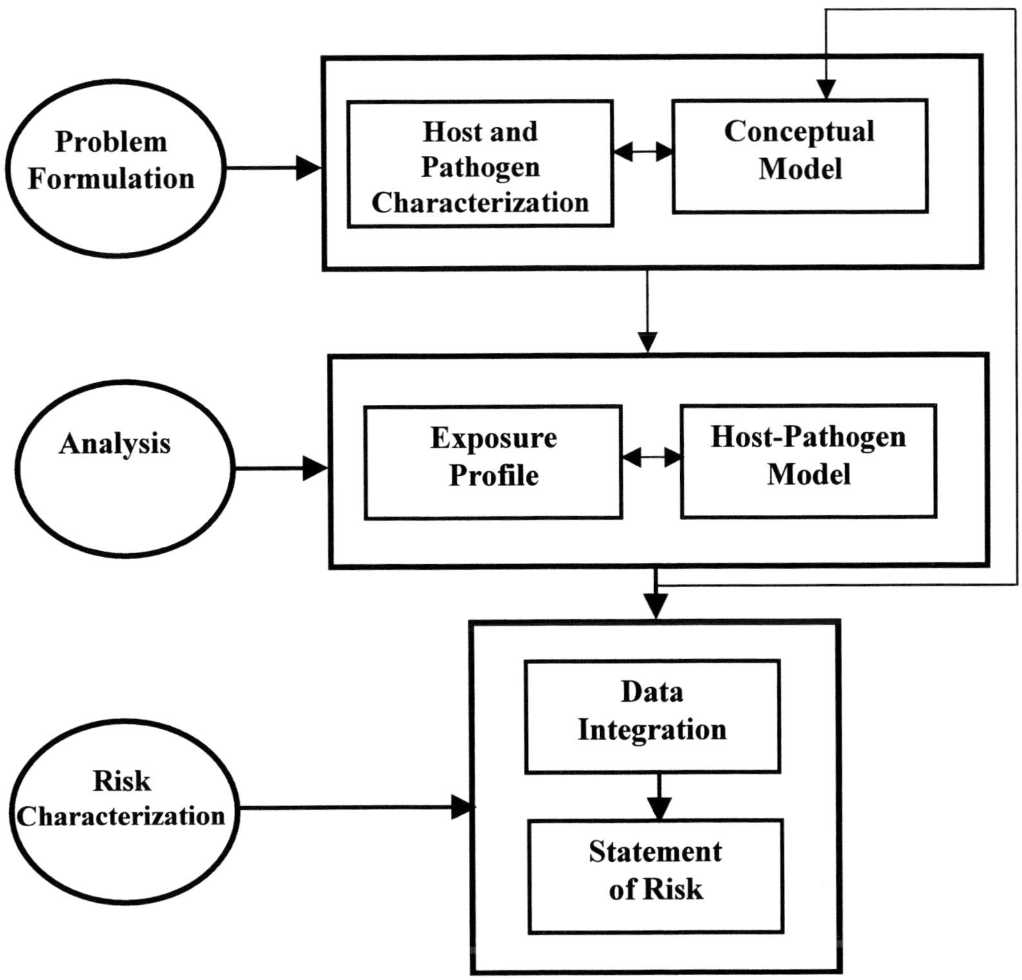

Figure 1-1. Schematic Overview Diagram of a Microbial Risk Assessment

Several important aspects of the microbial risk assessment process should be emphasized. First, iterative refinement is a fundamental component of the process. For example, data obtained during the problem formulation and refined during the analysis phase may be used to modify the conceptual models. The conceptual models may then be used to identify alternative sources of data, which may in turn be used to refine the model. Second, risk of infectious diseases is, by definition, manifested at the population level. Modeling of the infectious disease process, therefore, requires a dynamic model that can account for the unique features of disease transmission. In some cases it may be reasonable to make some simplifying assumptions to the disease transmission process. Several of the commonly applied assumptions in the field of microbial risk assessment are that each exposure event is independent, and person-to-person transmission of disease and immunity are negligible. If such assumptions are appropriate, a risk characterization approach based on the NRC chemical paradigm may reasonably be used to estimate risk. If any of those assumptions are not appropriate for a particular exposure-pathogen scenario, a more comprehensive risk characterization model would be necessary.

One such risk model is based on disease transmission. In this document, consistent with EPA microbial risk approach, the chemical risk model is extended to account for the unique properties of infectious disease processes using a disease transmission model. The principal approach taken in this research was to build upon a risk-based model that was previously developed for waterborne exposures (Eisenberg et al. 1996; Eisenberg et al. 1998; EOA 1999) and extend it to account for properties unique to biosolids and to exposure to pathogenic microorganisms from biosolids.

CHAPTER 2.0

MODEL DEVELOPMENT

Consistent with the EPA framework for microbial risk assessment, two of the fundamental components in the model developed for this investigation were exposure assessment and host-pathogen interaction (see Figure 1-1). The exposure assessment estimates the quantity of pathogens to which individuals or populations are exposed. This quantity is called the dose. Once calculated, dose is input into a disease transmission model, which in turn outputs a risk estimate associated with the specific pathogen-exposure scenario. This process is shown schematically in Figure 2-1. This chapter is divided into two sections. The first discusses the development of the exposure component, and the second describes the development of the health effects component.

2.1 Exposure Assessment

For the purposes of demonstrating the model, the exposure assessment focuses on one specific pathway, direct ingestion of biosolids amended soil. In the following sections, the rationale for its selection is discussed, data available to characterize that pathway are reviewed, and pertinent data related to human exposure to enteroviruses via a direct ingestion pathway are summarized.

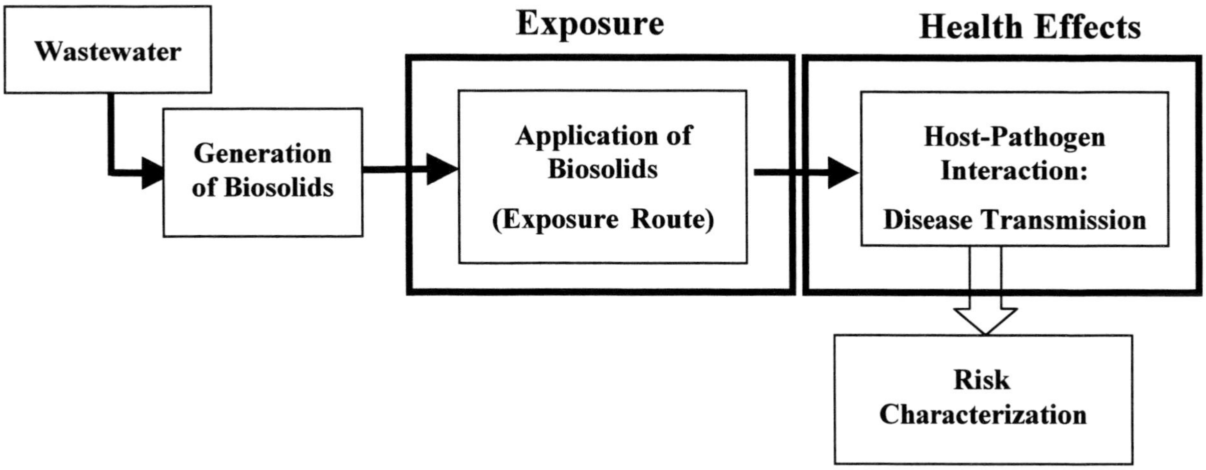

Figure 2-1. Risk Process Diagram

2.1.1 Biosolids as a Source of Human Exposure to Pathogens

The process of biosolids generation is shown schematically in Figure 2-2, which also outlines treatment processes that produce Class A and B biosolids. As shown in the figure, biosolids products are subject to environmental conditions prior to human exposure. Those environmental conditions vary temporally and from site to site, and therefore these conditions need to be understood in order to incorporate them into the model. The goal of the exposure assessment is to characterize the level of pathogens to which individuals may be exposed. The exposure assessment, therefore, is composed of characterizing two components: the concentration of the pathogen of interest at the time of exposure (# pathogens per unit mass or volume), and the mass or volume of the medium (e.g., mass of soil or volume of water) to which individuals or populations are exposed.

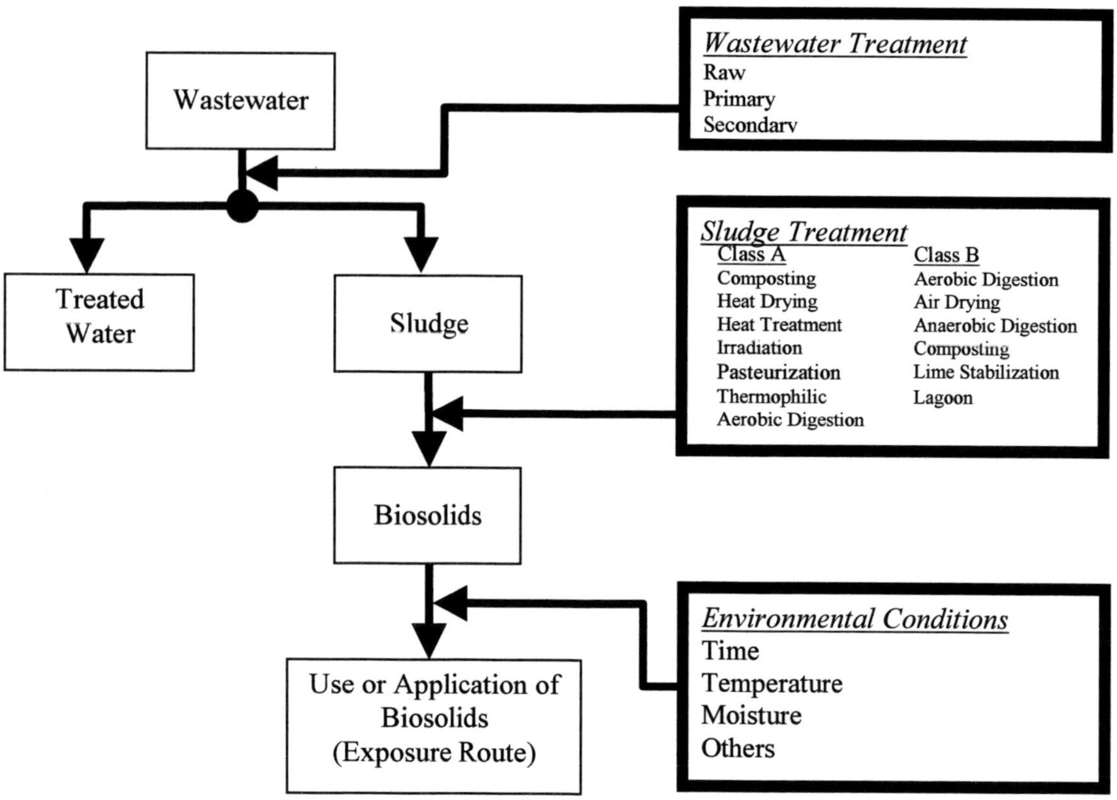

Figure 2-2. Risk Process Diagram for the Generation of Biosolids

Results from the 1989 EPA biosolids risk assessment, reviewed in Chapter 1, suggest that biosolids related exposures can be classified into one of five pathways. Those five pathways may be summarized as follows:

- Ingestion of surface water containing pollutants from eroded biosolids;
- Direct ingestion of biosolids-amended soil;

- Ingestion of groundwater contaminated via water leaching through treated soil;
- Ingestion of plants grown in biosolids amended soil; and
- Inhalation of pollutants in biosolids dispersed into the air.

Conceptual schematic models for these exposure pathways were developed for each of the exposure routes shown in Figure 2-3. These conceptual models outline the data required to quantify human exposure to pathogens from the beneficial reuse of biosolids (Appendix A). Once a pathway and pathogen are selected, the dose can be quantified and used as an input variable for a model that characterizes risk. The risk model used in this project is a disease transmission model.

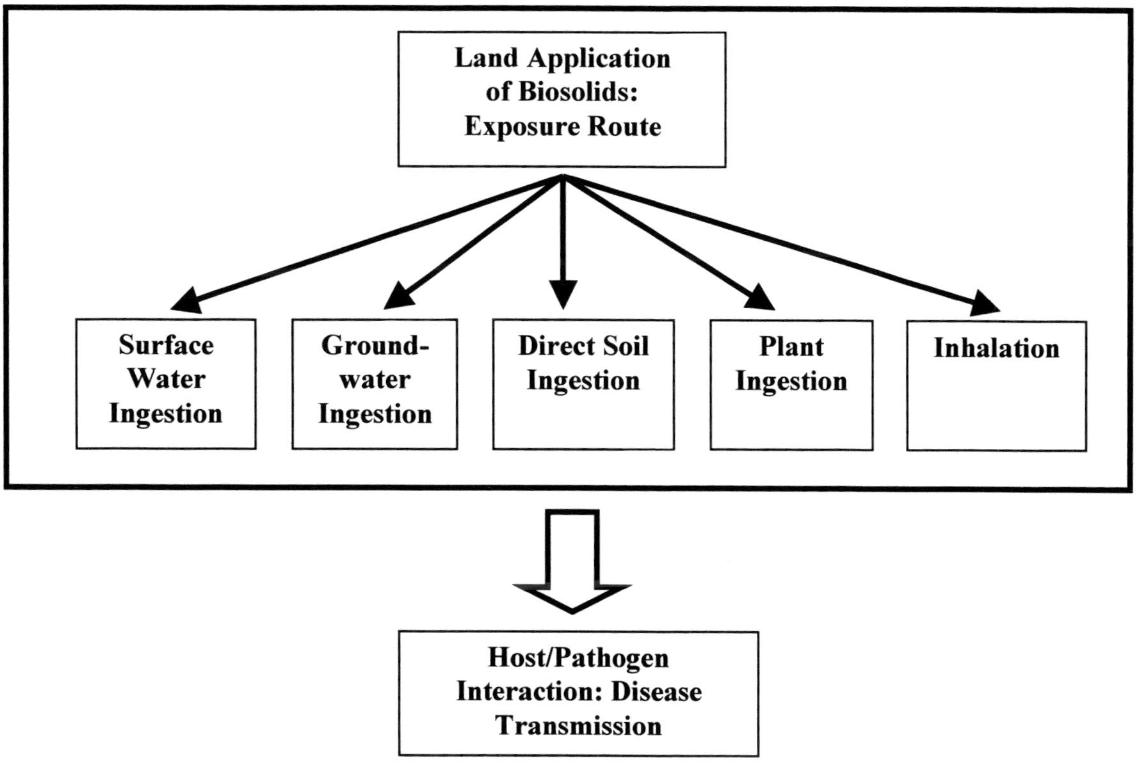

Figure 2-3. Relevant Routes of Exposure for Land Application of Biosolids

2.1.2 Selection of Route of Exposure and Pathogens

For the purposes of developing the risk model and demonstrating it in a transparent manner, a pathway that required a minimum number of assumptions was chosen. The direct-ingestion-of-soil pathway fit these criteria. It should be emphasized that the model described herein is applicable for all potential pathways of exposure to pathogens provided that the level of exposure can be characterized.

As an example, the conceptual schematic model for exposure to pathogenic organisms via the direct ingestion of biosolids-amended soil pathway is presented in Figure 2-4. A number of factors influence the level of exposure from ingestion of soil amended with biosolids, such as the application rate of the biosolids, the concentration of the pathogen of interest in the biosolids, additional sources of the pathogen to the soil prior to the time of ingestion, the die-off (or growth) rate of the pathogen in soil, the time between application of biosolids and ingestion of soil, and the mass of soil ingested.

A thorough understanding of this exposure model is not critical to an understanding of the overall risk model described herein. What is important is to appreciate that the output from the exposure assessment is a quantity (or a distribution) of pathogens that is required as input into the health effects model. For the example shown in Figure 2-4, the number of pathogens is derived as the concentration of pathogens in the soil at the time of soil ingestion times the mass of soil ingested.

The selection of representative pathogens for this investigation was the subject of much discussion with the PSC, HAC, and the research team. Based on discussions with the PSC and HAC, criteria were established for the selection of representative pathogens for the investigation. Those criteria were (1) Occurrence data must be available for the pathogen in biosolids; (2) Human dose-response data must be available for the pathogen; and (3) The pathogen must be of potential concern in biosolids.

Based on the criteria described above, the project team in agreement with the PSC and HAC decided that the following organisms would be investigated: enteroviruses, *Salmonella spp.,* and *Ascaris lumbricoides*. After a critical review of the literature, it was determined that the available human dose response data were not sufficient to carry out a quantitative evaluation for health effects associated with *Ascaris lumbricoides* in humans. The project team chose to focus on enterovirus rather than *Salmonella* because of the wider availability of data that were useful in identifying key model parameters. These data come from studies such as the New York and Seattle Viral Watch programs that followed households over time (Hall et al. 1970; Kogon et al. 1989).

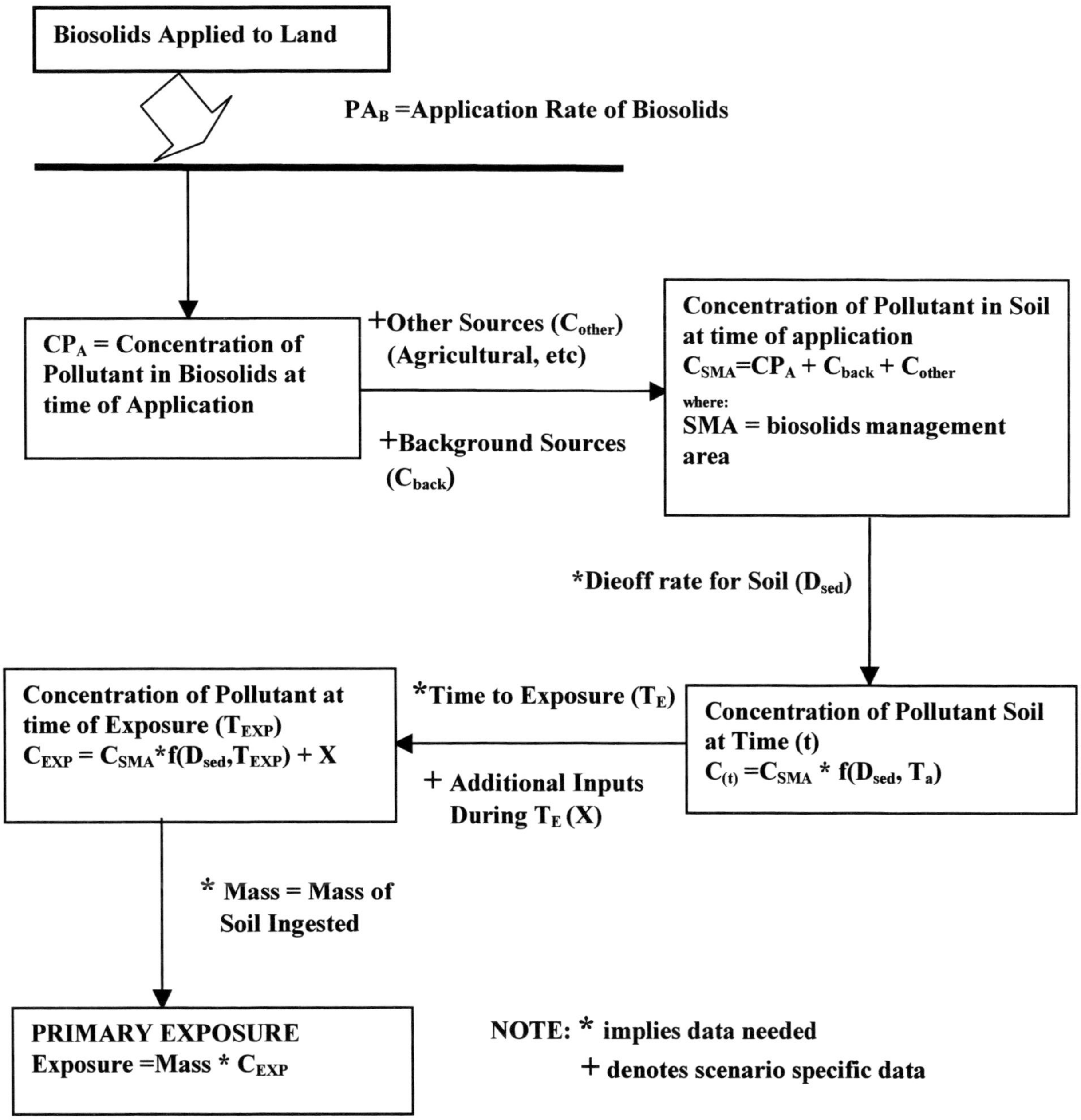

Figure 2-4. A Conceptual Model for Exposure to Pathogens through the Direct Ingestion of Soil

2.1.3 Characterization of Exposure: Data Requirements

Data needs for characterizing exposure are summarized schematically in Figure 2-5. The process of biosolids generation and subsequent human exposure is shown schematically in the center portion of the figure. The data that would be needed to characterize pathogen concentrations at each step along the process are shown on the left side of the figure.

The characterization of exposure to pathogens present in biosolids may be developed in numerous ways, depending on the available data. For example, occurrence data for a pathogen in fully treated biosolids, survival (and/or regrowth) rates for the pathogen during storage prior to and after application, and soil ingestion data are sufficient to characterize exposure. If occurrence data for the pathogen in fully treated biosolids were not available, occurrence data in (untreated, primary, or secondary) sludge could be substituted along with estimates of treatment efficiency for the processes used to generate the finished biosolids product. Uncertainty increases, however, as the data used in the exposure assessment is collected from locations that are further from the exposure site.

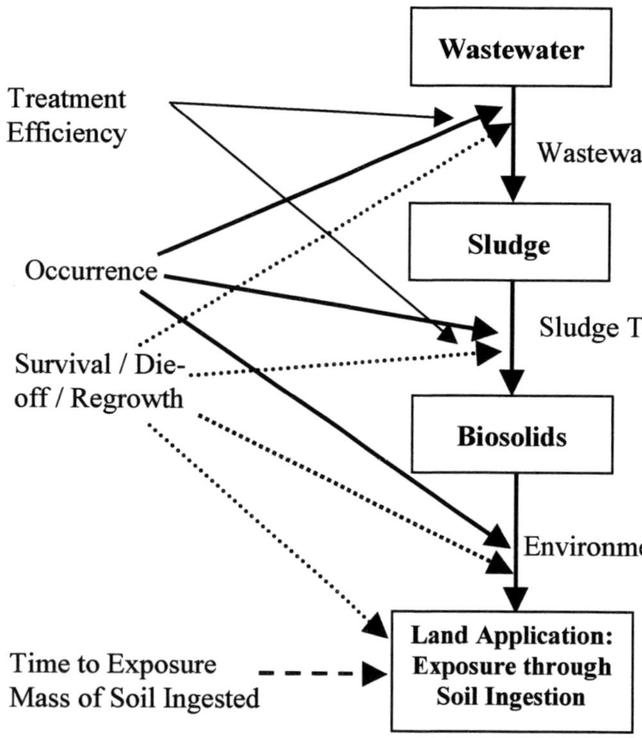

Figure 2-5. Data needs for characterizing exposure

2.1.4 Exposure Profile for Enteroviruses

The exposure profile provides information on the magnitude, frequency, and patterns of exposure for the scenarios to be used in the risk characterization. This information was obtained through a detailed literature review. Based on that review the most pertinent data were identified and condensed into an exposure profile. The following six categories were used to structure the exposure profile for enteroviruses:

1. Enterovirus occurrence in anaerobically digested biosolids;
2. Enterovirus occurrence in raw and secondary sludge;
3. Reduction of enteroviruses during mesophilic anaerobic digestion;

4. Storage of biosolids;
5. Survival in soil; and
6. Soil ingestion.

The data summarized in this profile were used to estimate parameter values for the exposure component of the disease transmission model. These parameter estimates are summarized in Table 3-2. Details of the exposure profile are provided in Appendix B.

2.2 Health Effects: Building the Conceptual Model for Disease Transmission

In Section 2.1, the exposure component was presented, followed by a review of available data to characterize the direct ingestion pathway for soil that has been amended with Class B biosolids. In this section, the other major component of a risk assessment, the health effects component, are presented. Figure 2-1 shows the link between the exposure and the health effects components of a risk assessment. The exposure assessment estimates a dose that is then used as the input variable for the disease transmission model. The disease transmission model in turn estimates a risk estimate. This risk characterization is conducted in the context of a specific pathogen/exposure scenario. Analogous to the exposure component, pathogen-specific health-effects information is required before conducting a risk characterization. This has traditionally involved identifying a dose-response relationship for the pathogen understudy. In Chapter 1.0, the need for additional pathogen-specific information for risk estimation is described. Other processes, such as secondary transmission and immunity potentially play an important role in estimating risks associated with exposure to pathogens present in the environment. In this chapter, a conceptual model for disease transmission is developed and the types of health-effects information needed to identify model parameters are discussed. At the end of this section there is a review of the relevant health effects data available for enteroviruses.

2.2.1 The Chemical Risk Model

The health effects component of a microbial risk assessment involves evaluating the likelihood that an adverse health effect may result from human exposure to one or more potential pathogens. The chemical risk model (NRC 1983) has commonly been used as a generic framework for carrying out microbial risk assessments related to water- and food-borne pathogens (Crabtree et al. 1997; Farber et al. 1996; Sanaa et al. 2000; Voysey and Brown 2000). These assessments have typically focused on estimating the probability of infection or disease to an individual as a result of a single exposure event. Moreover, these assessments have generally assumed that multiple exposures constitute separate and independent events with identical distributions of contamination, and secondary transmission and immunity within a population are negligible (Regli et al. 1991).

To demonstrate the link between a chemical risk model and a disease transmission risk model the basic structure of a microbial risk assessment is presented that employs the chemical risk model schematically in Figure 2-6. The two boxes labeled "Susceptible" and "Infected/Diseased" represent the possible disease states. The source of pathogens is shown in the dotted box labeled "Dose." In this model, pathogens can move from the environment to a susceptible individual (labeled "Environment-to-Person"). The solid line represents the movement of individuals from one disease state (e.g., susceptible) to another (e.g., infected).

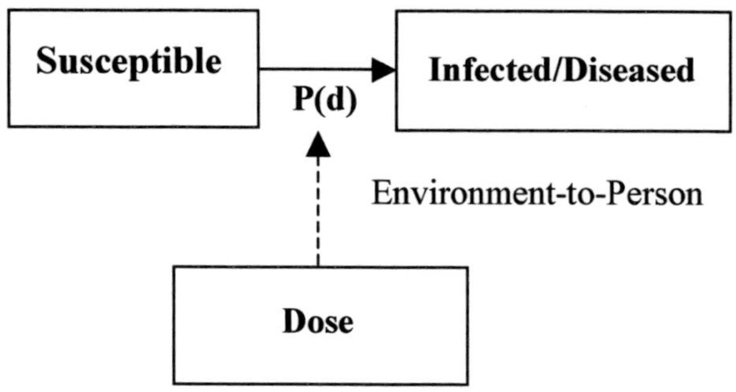

Figure 2-6. A Schematic Diagram of NRC Chemical Risk Model Applied to a Microbial Risk Assessment

The probability that a susceptible individual becomes infected or diseased is a function of the dose of pathogens to which they are exposed; i.e., when individuals in a susceptible state are exposed to pathogens from an environmental source, they move with a given probability to an infected or diseased state. This probability dose-response function is labeled P(d) in Figure 2-6. The dose is calculated by estimating two quantities, the concentration of pathogens at the exposure site and the ingestion rate of the susceptible individuals. This dose quantity is then input into a dose-response function that estimates an individual risk value.

The important health effects information required for the chemical risk model, therefore, is summarized in the function that represents this probability of infection, P(d), known as a pathogen-specific, dose-response function. The probability of infection following exposure to some virulent pathogen depends on several host- and pathogen-specific factors. The interaction between a pathogen and the host can be viewed as a series of conditional events, in which each must be achieved to result in infection. The infection status depends on a number of factors such as: (1) the number of organisms that enter the host; (2) the host's ability to inactivate these organisms; (3) the number of organisms that can withstand the host's local immune defenses, adhere to mucosal surfaces, and multiply in order to infect the host; and (4) variation in pathogen virulence and host susceptibility.

Infection is often defined as the invasion, colonization, and multiplication of a pathogenic microorganism (Teunis et al. 1996). Because humans may come into contact with a wide range of pathogen concentrations, it is useful to develop a functional relationship between a quantitative measure of microorganism ingestion and the risk of infection. A dose is defined as a quantitative measure for the intensity of exposure of the host to the pathogen of interest. The units of dose are usually given as the number of organisms ingested, but the units may vary from pathogen to pathogen. By studying the effects of various doses, a dose-response relationship can be determined between the dose and the frequency of infection within the exposed population of hosts. Quantitative descriptions of the dose-response relationships of many organisms have been studied in order to give insight into the risk of becoming infected after the ingestion of a certain dose of organisms (Teunis et al. 1996).

2.2.2 The Disease Transmission Paradigm

For both infectious and noninfectious disease processes, infectivity as a function of dose (estimated using a dose-response function) is an important factor to account for when estimating risk. The dose-response function is the critical component of a chemical risk assessment model. There are, however, other factors that also potentially play an important role in infectious disease processes such as person-to-person transmission, immunity, asymptomatic infection, and incubation period. Each of these is discussed in the sections that follow.

The potential for person-to-person transmission of disease is well documented in the literature for enteroviruses. For example, Melnick (1996) reported that enteroviruses can be excreted in stools for over a month. In addition, Melnick discusses how intrafamilial spread of non-polio enteroviruses is high and related to the duration of shedding. This rate of infection among non-immune family members may be as high as 80%. Consistent with the Melnick review, among the 93 members of the 19 Seattle Virus Watch families members who were exposed to aseptic meningitis in 1968, 73% became infected (Hall et al. 1970).

A review of the enteroviruses literature indicated that enterovirus-induced antibodies produced partial and temporary immunity to either infection or disease, but levels of protection vary from pathogen to pathogen. For example, it was observed in the New York Virus Watch program that household transmission of coxsackievirus was lower among exposed family members who had detectable neutralizing antibodies at the beginning of the study (25% if antibodies present versus 76% if no antibodies were present) (Kogon et al. 1969). Another illustration of protective immunity comes from an epidemic of acute meningitis in Seattle caused by echovirus 30. Data from the Seattle Virus Watch suggested that antibodies for echovirus 30 may have been protective, particularly for children. Among 18 families that acquired infection, only 3% of persons had antibodies. Among 46 families that did not acquire infection, 9% of persons had antibodies (Hall et al. 1970).

Infection may or may not be accompanied by symptoms. A person who has an asymptomatic infection (a carrier) may be of public health concern because the carrier may not be aware of his or her infection. Nevertheless, the carrier may continue to transmit the pathogen, contributing to its spread. The often-used serological test only signifies past exposure. To identify a carrier it is necessary to examine stool specimens for the presence of the pathogen. It is generally believed that the asymptomatic proportion of enterovirus infection is relatively high.

The definition of incubation period is the time required for multiplication of a parasitic organism within a host sufficient to produce symptoms. There has been much less work published on modeling incubation period than dose response, however Sartwell published three particularly relevant articles on incubation period (Sartwell 1950; Sartwell 1952; Sartwell 1966).

In his first paper Sartwell fit incubation period data for a wide range of pathogens to lognormal distributions. One important finding by Sartwell relevant to model development is the observation that dispersion factors do not significantly vary across the pathogens studied and are independent of the mean length of incubation. This finding, coupled with a parametric assumption of how incubation periods are distributed, allows us to use data that only contain median incubation periods.

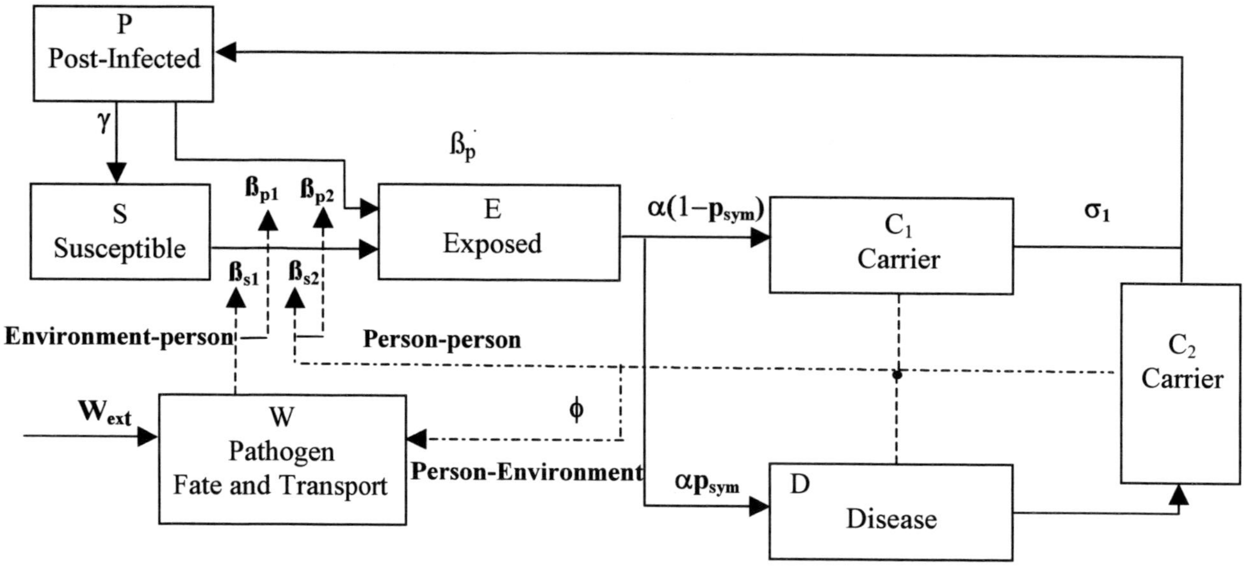

Figure 2-7. Disease transmission model.

2.2.3 Model Development

The previous section discussed factors not accounted for in the chemical risk paradigm, such as person-to-person transmission and immunity. To account for these additional factors when estimating risks associated with exposure to pathogenic microorganisms, a more sophisticated mathematical model than the one shown schematically in Figure 2-6 is required. In this section, a conceptual disease transmission model is developed that is capable of accounting for the attributes discussed above.

The model shown in Figure 2-6 was extended to account for person-to-person transmission, immunity, incubation, and asymptomatic infection. These factors are shown schematically in Figure 2-7.

The model shown in Figure 2-7 contains six disease states as compared to the two disease states presented in Figure 2-6. These six disease states are: (1) the susceptible state, *S*, defined as those who are susceptible to disease; (2) the exposed state, *E*, defined as those who have been exposed to pathogens but are asymptomatic and noninfectious; (3) the carrier state 1, C_1, defined as those who are infectious carriers but are asymptomatic; (4) the diseased state, *D*, defined as those who are symptomatic and infectious;(5) the carrier state 2, C_2, defined as those who were previously symptomatic but are now asymptomatic and infectious; and (6) the protected state, *P*, defined as those in a postinfectious state who are noninfectious and exhibit some level of immunity. Similar to Figure 2-6, the solid lines represent the movement of individuals from one disease state to another.

The rates of movement of individuals from one disease state to another are parameters represented in Figure 2-7 by Greek letters. For example five of these parameters represent processes that occur postinfection: (1) the rate of movement from the exposed state to an infectious state is represented by α, where $1/\alpha$ corresponds to the latency period; (2) the rate of movement from the symptomatic infectious state to the asymptomatic infectious state is represented by δ, where $1/\delta$ corresponds to the duration of symptoms during infection; (3) the rate of movement from the asymptomatic infectious states to the post-infection state is represented by both σ_1 and σ_2, where $1/\sigma_1$ and $1/\sigma_2$ correspond to the duration of infectiousness; and (4) the rate of movement from the postinfection state to the susceptible state is represented by γ, where $1/\gamma$ corresponds to the duration of immunity. The reciprocal rates correspond to the duration in which an individual is in one of the following states: latently infected (E), asymptomatically infectious (C_1 and C_2), symptomatically infectious (D), or partially protected from further infection (P). These states can therefore be modeled mathematically as delay processes (see Appendix C for details). An additional two parameters complete the postinfection list of factors: p_{sym}, the proportion of symptomatic infections; and ϕ, the rate that infectious individuals shed pathogens into the environment.

In contrast, the transmission rate parameters, β_{s1}, β_{s2}, β_{p1}, β_{p2}, represent the infection event. The first subscript, *s* or *p*, represents an infection event of a susceptible or partially protected individual. The second subscript, *1* or *2*, represents an infection event through a different transmission route, either through the environment (subscript 1) or directly person-to-person (subscript 2). These transmission routes are represented by dotted lines in Figure 2-7. In one of the transmission routes, pathogens may move from infected individual to susceptible individual (labeled "person-person transmission"). This transmission route is described by β_{s2} and can be thought of as a product of two terms: a rate of contact between an infectious and susceptible individual and a probability of infection given that a contact occurs. The parameter β_{p2} is analogous to β_{s2} except the transmission event occurs in an individual who is partially protected. Initially, individuals in state P are completely protected, i.e., β_{p2} is equal to zero times β_{s2}. The protection decreases linearly in time until the duration of protection has ended at which point $\beta_{p2} = \beta_{s2}$ and individuals transition from the P (partially protected) state to the S (susceptible) state (see Appendix C for details).

In the other transmission route, pathogens may move from an infectious individual to the environment and subsequently from the environment to a susceptible individual. (labeled person-to-environment-to-person transmission). This environmental transmission route accounts for infected individuals shedding pathogens into the waste stream. The parameter ϕ characterizes the intensity of shedding from infectious individuals. These pathogens may eventually circulate back into the community through an environmental pathway (for example, through discharge of a wastewater treatment plant or via the generation and reuse of biosolids). This component accounts for the fate and transport of pathogens in the environment, as depicted in Figure 2-7 and as shown in more detail in Figure 2-8.

As shown in Figure 2-7, pathogens enter the environment either via infected individuals (persons from state C_1, C_2, or D) who shed ϕ pathogens per person per day, or via some external source, represented by W_{ext}. Pathogens remain in the system until they die off naturally, are eliminated by treatment or some other environmental process, or infect a susceptible or partially susceptible human host. A distributed delay was used to characterize the average time pathogens spend in the environment before being processed in a medium such as biosolids. This fate and transport process is characterized by two parameters, k_w, the rate that pathogens move from contamination site to exposure site; and μ, the natural mortality rate during this process (see Appendix C for details). A fraction, f, of the pathogens is processed as biosolids and is, therefore, eliminated through biosolids processing. The parameter X_b represents this attenuation. Those pathogens not in biosolids, the fraction 1-*f*, are processed through other means such as drinking water treatment. The attenuation associated with these processes is represented by X_e.

In the example here, individuals are exposed to pathogens via biosolids by ingesting soil treated with biosolids. Exposure via an environmental route (not biosolids) also occurs through ingestion, where i_b and i_e are the biosolids and environmental ingestion rates respectively. This process results in a dose that an individual is potentially exposed and is estimated in this model as:

$$d = W[fX_b i_b + (1-f)X_e i_e] \tag{2.2.3-1}$$

where W is the concentration of pathogens coming from the wastewater stream.

A beta-Poisson dose-response function was used to determine the probability that an exposed person would become infected given a specific dose (Teunis et al. 1996):

$$P = 1-(1+\frac{d}{B})^a \tag{2.2.3-2}$$

where P represents the probability that an exposed individual will become infected, d represents viral dose, and a and B are the parameters of the model that are identified when the function is fit to dose-response data. Once infected, individuals move from the susceptible state to the exposed state. The rate of this movement is defined as β_{s1} to distinguish the rate of environmental transmission from secondary transmission (person-to-person). The rate of movement of individuals from a partially protected state to the exposed statc is similarly denoted by β_{p1} and is related to β_{s1} in the same manner as β_{p2} is to β_{s2}.

A summary of the components contained within Figures 2-7 and 2-8 is provided in Table 2-1 and mathematical equations for implementing the model are presented in Appendix C.

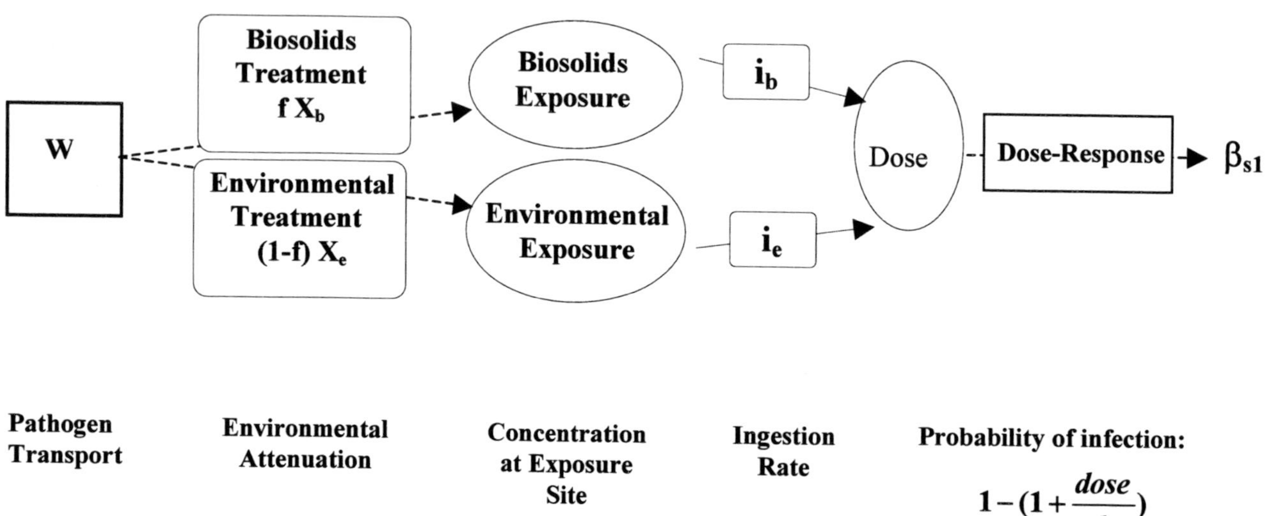

Figure 2-8. Pathogen Fate and Transport (Environmental Component of Transmission Model)

Table 2-1. Summary of Components in Basic Conceptual Model for Disease Transmission

Epidemiological States		Description
S		Number of individuals in susceptible state.
E		Number of individuals who have been exposed.
C1		Number of individuals in carrier state, did not experience symptoms.
D		Number of individuals in diseased state.
C2		Number of individuals in carrier state, did experience symptoms.
P		Number of individuals in postinfection state.
Rate Parameters		
α		Transition rate out of exposed state. Inverse of incubation period.
P_{sym}		Probability of symptomatic response.
δ		Transition rate out of diseased state. Inverse of disease duration.
σ		Transition rate out of carrier state. Inverse of duration of shedding.
γ		Transition rate out of postinfection state. Inverse of duration of immunity.
β	β_{s1}	Transmission rate (susceptible to exposed from environmental exposure)
	β_{s2}	Transmission rate (susceptible to exposed from person-to-person contact)
	β_{p1}	Transmission rate (postinfection to exposed from environmental exposure)
	β_{p2}	Transmission rate (postinfection to exposed from person-to-person contact)
Environmental Parameters		
ϕ		Rate that infectious individuals shed into the environment.
f		Fraction of the pathogens that are processed as biosolids.
X_b		Attenuation of pathogens through biosolids processing.
X_e		Attenuation of pathogens through environmental processing (other than biosolids).
i_b		Ingestion rate through biosolids exposure.
i_e		Integestion rate through general environmental exposure.
k_w		Rate of pathogen movement from contamination to exposure site.
μ		Pathogen die-off rate.
Input Variables		
W_{ext}		External source of pathogens.

2.2.4 Health Effects Profile for Enterovirus

The health effects profile provides the data needed to identify the parameters of this disease transmission model. Similar to the exposure profile, a detailed literature review was carried out to identify appropriate data for the health effects component of the risk characterization. Based on this review, the most pertinent data were identified and condensed into a health effects profile.

The following seven categories were used to structure the health effects profile for enteroviruses:

1. *Epidemiology of enteroviruses:* It has been reported that there are 20 million enteroviral infections per year in the United States (Pallansch 1997). The common enteroviruses are divided into three major groups of serotypes: coxsackieviruses; echoviruses; and enteroviruses;
2. *Probability of illness:* In general, data in the literature suggest that approximately half of enteroviral infections are asymptomatic (Melnick 1996);
3. *Immunity and seroprevalence:* The presence of serum antibodies was a strong indicator of prior exposure and some degree of protection to further infection;

4. *Shedding:* Enteroviruses can be excreted in stools for more than a month, but shedding from pharynx/upper alimentary tract generally lasts 1 to 2 weeks after infection (Melnick 1996);
5. *Person-to-person transmission:* It has been reported that intrafamilial spread of non-polio enterovirus (NPEV) is high and is related to the duration of shedding (Melnick 1996);
6. *Dose response:* Dose response is species and serotype specific for enteroviruses. Dose response investigations have been carried for three of the enteroviruses: coxsackie A21; echovirus 11; and echovirus 12;
7. *Incubation:* The incubation rate for enteroviruses ranges from 2–20 days but is usually 5–14 days (Melnick 1996).

The data from this literature search were used to estimate the parameter values of the disease transmission model. The parameter estimates are summarized in Table 3-1. Details of the health effects profile are provided in Appendix D.

CHAPTER 3.0

MODEL DEMONSTRATION: CASE STUDY AND SENSITIVITY ANALYSIS

The value of a risk assessment resides in its ability to provide information to the risk manager or regulator. Typically, one goal of a risk assessment is to provide a risk estimate for a particular exposure scenario and a given pathogen. This risk estimate can take on the form of a point estimate, e.g., the individual annual risk of exposure to *Giardia* from drinking water exposures is 3.4×10^{-6}, or it can take on the form of a distribution, e.g., there is a 65% probability that the risk is between 1.0×10^{-7} and 1.0×10^{-5}. One motivation for using a distribution to describe risk is that exposure and dose-response measurements are highly uncertain. In addition, pathogen and exposure levels, even if accurately measured, are highly variable. Risk to an individual or population, therefore, cannot be appropriately estimated using a single value. It is best characterized as a distribution of values with a mean and standard distribution.

The perspective taken in this report is that risk assessments can generate more than simple risk estimates. They can also be used as a comprehensive tool to explore alternative scenarios for risk management. A risk manager may want to know the benefits associated with modifying a given component within a treatment process. In this case the risk methodology described in Chapter 2.0 can be used to estimate these benefits in terms of disease reduction. An example would be to use the model to estimate the reduction in risk associated with increasing the storage time of biosolids processing. Similarly, a regulator may want to know the benefits associated with implementing a given rule. For the regulator, the risk methodology can provide an estimate of the benefits, in terms of the disease reduction associated with the lower contaminant level.

In addition to providing valuable information to the risk managers and regulators, the risk model provides information on data needs. Through sensitivity and uncertainty studies, parameters that are most responsible for uncertain risk estimates can be identified. As a result, information from targeted research and data collection efforts can be used to reduce risk uncertainties. These data needs reflect the additional information that would provide the largest decrease in the uncertainty of the risk estimate. Alternatively, the risk model can be used to increase the confidence in a given policy decision. In this context, rather than looking at ways to minimize the uncertainty in the risk estimate, sensitivity studies can be used to examine the probability that added information will affect a given decision under consideration.

In this chapter the features of the risk methodology is presented. This demonstration is not an actual risk assessment, but rather a simulation study that examines the interactions of factors that govern the risk process.

3.1 Methods

3.1.1 Scenario Definition

Risk characterizations are generally conducted for site-specific situations in which the environmental and behavioral conditions are well defined. To demonstrate this methodology, therefore, a scenario was developed that had sufficient degree of realism to define these environmental and behavioral conditions. This scenario, although realistic, was not meant to account for all types of biosolids exposures. Rather, it was simply meant to provide a construct from which to illustrate the utility of the approach.

The community characterized in this scenario was assumed to be closed to external sources of pathogens; that is, infectious community members generated pathogens in the environment. The community members were assumed to be homogeneous with respect to exposure and susceptibility to disease. With respect to the pathogen, enteroviruses were assumed to be endemic in the community and the only source of enterovirus pathogens in the environment came from infectious individuals. In addition, the virus population was assumed to be homogeneous; distinct characteristics of the different enterovirus strains were not modeled.

Pathogens were assumed to be transported to the biosolids exposure site via either the wastewater stream from within the community, or from an external source from outside of the community. Through this transport the pathogens were attenuated by a wastewater treatment facility, natural die-off, and dilution. Upon reaching the biosolids processing stage it was assumed that Class B biosolids were generated; that is, wastewater sludge underwent mesophilic anaerobic digestion for 20 days at 35°C and was then dewatered to 20% solids and air dried for 3 months.

With regard to application, it was assumed that biosolids were applied only once and that no other biosolid exposure sources existed; i.e., community members could only be exposed to biosolids by ingesting treated soil. The assumption in the simulations is that the population is homogeneous; i.e., subpopulation groups are not modeled. The units of the state variables were in fractions; i.e., $S = 0.4$ reflects that 40% of the population is in the susceptible state. An estimate of the population size is, therefore, not required.

3.1.2 Parameterization

The parameter values used in the simulation study are shown in Tables 3-1 and 3-2. A low, medium, and high value was assigned for each parameter based on information presented in the literature. If no information was available, appropriate assumptions were made as specified in the comments column in the table.

The 8 parameters in Table 3-1 are related to the disease process of the pathogen within the human host. The first five parameters represent the postinfection processes described in Section 2.2.3. Data used to establish these ranges were found in the literature as detailed in

Appendix E. For example, the incubation period was assumed to be one of three values — 2, 10, or 40 days — corresponding to a rate for α of 0.2, 0.4, or 2.0 per day. The middle value was intended to represent a more typical value, whereas the high and low values are representative of more extreme but plausible values. Similarly, data on the duration of symptoms and shedding was used to set values for δ, σ_1, and σ_2. The values for the rate of leaving the postinfection stage, γ, come from a variety of studies described in the previous chapter that examined the duration of protection from prior exposure. Three parameters related to the infection processes: the proportion of infections that are symptomatic, p_{sym}, and shedding rate of infectious individuals, ϕ, as well as the person-to-person transmission rate parameters, β_{s2} are also shown in Table 3-1. Whereas the values for p_{sym} were based on information from the literature, the values for ϕ and β_{s1} were chosen to explore the behaviors of the model. In this manner the simulation analysis was designed to identify the conditions for which the different transmission processes were important in determining risk.

Table 3-1. Summary of parameters associated with the disease process for enterovirus See Appendix C for information on the sources used to obtain these values

	Parameter Description	Low estimate	Middle estimate	High estimate
α	Transition rate out of exposed state. Inverse of incubation period.	0.2	0.4	2
δ	Transition rate out of diseased state. Inverse of disease duration.	0.2	0.8	4
σ_1	Transition rate out of carrier state. Inverse of duration of shedding.	0.066	0.4	2
σ_2	Transition rate out of carrier state. Inverse of duration of shedding.	0.066	0.4	2
γ	Transition rate out of post infection state. Inverse of duration of immunity.	0.008	0.04	0.4
P_{sym}	Probability of symptomatic response.	0.25	0.5	0.75
ϕ	Intensity of shedding.	10^4	10^6	10^8
β_{s2}	Transmission rate (susceptible to exposed from person-to-person contact)	0.01	0.05	0.1
β_{p2}	Transmission rate (partially protected to exposed from person-to-person contact)			

The remaining parameters are environmental factors that are all summarized into a dose value that is in turn used to identify the environmental transmission rate, β_{s1}. The first two parameters shown in Table 3-2 represent external sources of pathogens. These parameters are set to zero for this case study, since, as previously mentioned, the scenario chosen was a closed community in which all pathogens in the environment come from the community. The next two parameters pertain to the fate and transport of the pathogen from contamination site to exposure type. The values chosen were set to explore a range of possible transport times, from 6 to 30 days, as well as a range of mortality rates that correspond to average life spans from 6 to 30 days.

The next seven parameters are required to estimate the probability of infection from the environment, where the dose that an individual is exposed to is estimated to be

$$d = W[fX_b i_b + (1-f)X_e i_e] \tag{3.1.2-1}$$

where W is the concentration of pathogens coming from the wastewater stream, f is the proportion of pathogens that enter the biosolids stream, X_b is the attenuation of pathogen concentration caused by the biosolids process, X_e is the attenuation of pathogen concentration caused by all the other environmental pathways besides biosolids, and i_b and i_e are the biosolids and environmental ingestion rates respectively. The biosolids attenuation parameter, X_b, was modeled in more detail as the product of the last six parameters shown in Table 3-2. These values were chosen using information from the literature (Appendix E). Likewise, data from the literature was used to identify the two ingestion rate parameters. Little is known, however, about the remaining two parameters, f and X_e. For this reason, parameter values were chosen that would allow for the exploration of their role in affecting risk estimates.

As discussed in Chapter 2.0, the dose is then used as an input variable of a dose-response function, which in turn is used to estimate β_{s1}. The dose-response function is of the following form:

$$\beta_{s1} = 1-(1-\frac{d}{B})^{a} \qquad (3.1.2\text{-}2)$$

Table 3-2. Summary of Parameters Used to Estimate the Environmental Transmission Parameter, β_{s1} (See Appendix C for information on the sources used to obtain these values)

	Parameter Description	Low estimate	Middle estimate	High estimate
W_{extb}	Concentration of virus in sludge, from external source	0	0	0
W_{exte}	External virus from environment	0	0	0
ν	Pathogen die-off	0.033	0.066	0.166
k_w	Rate of movement of organisms from contamination site to exposure site.	0.133	0.266	0.666
f	% pathogens from W present in biosolids	0.1	0.25	0.5
i_b	Ingestion rate of treated soil (g/day)	0.001	0.01	0.2
ie	Ingestion rate though environmental routes (g/day)	2	2	2
a	Dose-response parameter	0.126	0.26	0.5
B	Dose-response parameter	.21	.42	.84
X_e	Environmental attenuation factor	0.0001	0.01	0.1
X_b	Biosolids attenuation factor (computed as the product of the 6 parameters listed below)	0.0001	0.01	0.1
r	Treatment reduction. Log removal due to treatment.	0.01	0.05	0.1
s	Storage reduction rate. Reduction rate due to storage of biosolids prior to application	0.9985	0.9988	0.9992
t_s	Time of storage prior to land application (days)	90	90	90
a	Application reduction. Reduction factor to account for application methods	1	1	1
1	Die-off rate	0.1	0.237	0.562
t_e	Time to exposure (days)	365	7	1

3.1.3 Measure of Risk

Cumulative incidence (CI) was used to measure disease occurrence or morbidity. Cumulative incidence was defined as the number of new cases of disease during a 1-year period divided by the total population at risk. For example, if a population of 1000 people is at risk of developing disease, and 15 people actually develop it over the time span of one year, the resulting CI would be 15/1000. This is analogous to the risk number used in more traditional risk assessments and can be interpreted as an annual per-person risk of 0.015. The CI measure can therefore be interpreted as a risk estimate. In the model, the cumulative incidence was calculated by summing (or integrating) the number of new cases entering into state *D* at every time step. This integration was calculated over one year. The cumulative incidence was represented by the following equation:

$$CI = \int_{t=0}^{365} p_{sym} \cdot E \cdot dt \tag{3.1.3-1}$$

For each parameter combination, two simulations were run. The first was to simulate the condition in which biosolids exposure was present (CI_b). The second was to simulate the condition in which biosolids exposure was absent (CI_o). This second condition was accomplished by setting the parameter X_b equal to zero. To calculate the number of cases that were due to biosolids, CI_0 was subtracted from CI_b. This difference in disease incidence between exposed and unexposed groups is attributable risk. Thus, the attributable risk due to biosolids (AR_b) is defined as the following:

$$AR_b = CI_b - CI_o \tag{3.1.3-2}$$

To illustrate this risk measure, two examples are presented, one illustrating a condition in which AR_b is low and one illustrating a condition in which AR_b is high. In the first example, parameter values were chosen to simulate a condition in which biosolids exposure was relatively low and secondary transmission was high. When simulated with biosolids present the model produced a one-year cumulative incidence of 2.61 cases per person per year; i.e., on average, each individual is infected 2.61 times per year. When simulated with biosolids absent the model produced a one-year cumulative incidence of 2.58 per person per year. Therefore the attributable risk associated with biosolids in this hypothetical example was $AR_b = 0.03$ cases per person per year. The fact that the baseline incidence of the pathogen was high suggests that in this hypothetical example the pathogen was highly infectious and is able to move through the population effectively through person-to-person transmission. The fact that the attributable risk was low suggests that only a small fraction of those infections are due to actual biosolids exposure.

In an alternative example, parameter values were chosen to simulate a condition in which the exposure to biosolids was high and secondary transmission was low. The results of the model simulation produced a 1-year cumulative incidence when biosolids were present equal to 0.83 cases per year per person and a cumulative incidence of 0.33 per person per year when biosolids were absent. Therefore, the attributable risk associated with biosolids in this hypothetical example was $AR_b = 0.5$ cases per person per year.

This attributable risk measure is equivalent to that used in a more traditional analysis (one that ignores person-to-person transmission). Using the chemical risk paradigm, concentrations of the pathogen at the exposure site are measured and then used to estimate a dose. A dose-response function is used to estimate the risk associated with the exposure to biosolids. This risk estimate accounts for the direct risks due to ingestion of a given dose of pathogens. Using a disease transmission model the risk associated with biosolids exposure is also estimated. The difference is that this risk is based on not only the direct risk associated with an individual ingesting pathogens from the environment, but is also based on the indirect risks that are due to the propagation of the pathogen through person-to-person transmission.

3.1.4 Simulation Approach

3.1.4.1 Initial Conditions

At the beginning of each simulation, 95% of the population was placed in the susceptible state, and the remaining 5% was placed in the exposed state. Each simulation was run for 1500 days to ensure that steady state was reached; i.e., the number of people in each state reached equilibrium. After 1500 days had elapsed, the number of disease onsets for a span of 1 year was counted. The risk estimate, therefore, corresponded to a 1-year incidence associated with enterovirus disease in an endemic setting.

3.1.4.2 Sensitivity Analysis

A sensitivity analysis was conducted to determine which parameters in the model impacted AR_b the most; i.e., to determine the parameters to which the output was most sensitive. Conducting a sensitivity analysis with all 22 parameters would require 3^{22} simulations, an unmanageable amount of computing time. Based on preliminary simulation, the output showed no evidence of being sensitive to a number of parameters, such as the rate of moving from the infected to infectious state. The six parameters that comprise the attenuation parameter, X_b, also were not included separately in analysis. When a group of parameters are expressed in a model as a product and never individually, one can only analytically derive inference from that product. Finally, some simplifying assumptions were made, such as setting the two duration-of-shedding parameters equal to each other. As a result of these simplifications, the following 10 parameters were included in the sensitivity analysis:

$\mathbf{X_b}$	Attenuation of pathogens in biosolids
$\mathbf{X_e}$	Attenuation of pathogens in the environment (other than biosolids)
f	Fraction of environmental pathogens that are in biosolids
σ	Rate of loss of infectiousness
α	Rate of moving from infected to infectious
β_{s1}	Secondary transmission rate
B	Dose response parameter
a	Dose response parameter
γ	Rate of decline in immunity
ϕ	Shedding rate.

Each parameter was assigned a high, medium, or low value from the available literature (see Tables 3-1 and 3-2). Parameters not involved in the analysis were assigned a medium value.

Simulations were conducted for each possible combination of parameters. This resulted in a total of 3^{10} (59,049) simulation runs. In each case AR_b was calculated. Thus, each unique set of parameters had a corresponding AR_b. A classification and regression tree analysis (CART) was used to classify the outputs into low and high AR_b. The results of the CART analysis were summarized in a tree diagram that was used to prioritize the importance of the model parameters in determining the conditions under which AR_b was high or low.

The CART analysis produces a tree structure. Each point where the tree branches is a node. At each node a parametric decision is made by an inequality $p_i < c$, where p_i is the ith parameter and c is a specific value of that parameter (Breiman et al. 1984). Observations that satisfy this condition are sent to the left node; otherwise they are sent to the right node. Each node is classified as a 1 if the number of 1 classifications (those that are classified as a high attributable risk) is greater than the number of 0 classifications (those that are classified as a low attributable risk). Otherwise, the node is classified as a 0. An example of a parametric decision is that the incubation period is less than 8 days, which corresponds to $\alpha < 0.125\ d^{-1}$. If out of 1000 simulations, 500 used an α that was less than $0.125\ d^{-1}$, and of those 500 simulations 400 corresponded to a low attributable risk, then that node was classified as a 1.

CART uses the minimization of the classification error as the criterion to decide which parameters are used to create the nodes. The number of classification errors, ε, for that node is then either the number of 1 classifications if the node was classified as a 0, or the number of 0 classifications if the node was classified as a 1. In this example, the node was classified as a low attributable risk condition. However, 100 of the 500 simulations had a high attributable risk. Therefore, in this example, there were 100 misclassifications.

The classification error for a given tree is the sum of misclassifications over every node divided by the total number of observations.

$$\frac{\sum_{m} \varepsilon_i}{N} \qquad (3.1.4.2\text{-}1)$$

where m is the total number of terminal nodes (nodes in which no subsequent splitting occurs), ε_i is the classification error at node i, and N is the total number of observations. When a specific split is chosen that minimizes the classification error, the node is replaced by two daughter nodes. In general as a tree size increases the error is reduced, and, as the tree size decreases the error is increased. The objective is to find an optimal combination of tree size and error. See Breiman (1984) for details.

3.2 Results

Analysis of the simulations identified four factors as the most important in determining the level of risk attributable to biosolids (AR_b). The first factor of importance was the relative contribution of biosolids toward exposure, compared to all other environmental pathways. In this model, this is represented as the ratio of X_e to X_b. The parameter X_b represents the percent reduction in pathogens attributable to a variety of biosolids-related processes and X_e represents pathogen reduction for all other environmental sources other than from biosolids; for example, an X_b value of .01 indicates a two-log reduction of pathogens during biosolids processing whereas an X_b value of 1 indicates no reduction at all. The parameter f represents the fraction of all pathogens in the environment that enter into the biosolids stream (Figure 2-8). If, for a given value of f, X_b is small (where smaller values correspond to greater attenuation) compared to X_e then a greater proportion of pathogens come from environmental pathways other than biosolids. Under these conditions, the biosolids pathway has a minimal contribution to the disease incidence and therefore the attributable risk associated with biosolids (AR_b) is small. The relative contribution of biosolids also depends on f. This relationship comes from comparing the contribution of biosolids to the total dose of pathogens that an individual is exposed to, WfX_b, with the total dose of pathogens associated with other environmental pathways, $W(1\text{-}f)X_e$. The relative contribution of biosolids can be then summarized by

$$\frac{fX_b}{(1-f)X_e} \tag{3.2-1}$$

Second in importance was rate of shedding from infectious individuals. Rate of shedding, ϕ, was defined as the number of pathogens shed per person per day. Only moderate values of ϕ resulted in an increased AR_b. When shedding was very high or very low, biosolids were not a factor in disease incidence.

The rate of person-to-person (secondary) transmission was a third factor impacting AR_b. This was represented by β_{s2} in the model. When β_{s2} was high, person-to-person transmission accounted for a greater proportion of infections. Under this condition, competing transmission pathways, such as those associated with biosolids and environmental exposures, account for fewer infections. As a result, AR_b was low.

The fourth factor found to affect AR_b was immunity. Movement from the protected state to the susceptible state was governed by γ. Duration of immunity was inversely proportional to γ; i.e. for small values of γ, duration of immunity is high. Shorter periods of immunity led to higher values for AR_b while longer periods of immunity led to smaller AR_b values.

The hierarchy of importance exhibited by these factors is important. As can be seen in Figure 3-1, the impact of each successive factor is conditional on the existence of the preceding factor; i.e., secondary transmission influences AR_b but only if shedding is moderate and the relative contribution of biosolids towards exposure is high. In fact, if any of the following conditions were true (1) the contribution of biosolids towards exposure was small compared to the contribution from other sources; (2) the rate of pathogen shedding was relatively high or low (not moderate); (3) person-to-person transmission was relatively high, then AR_b was low regardless of other parameter values.

The resulting CART tree from the simulation studies is shown in Appendix F. The following four sections describe these parameters and their impact on risk.

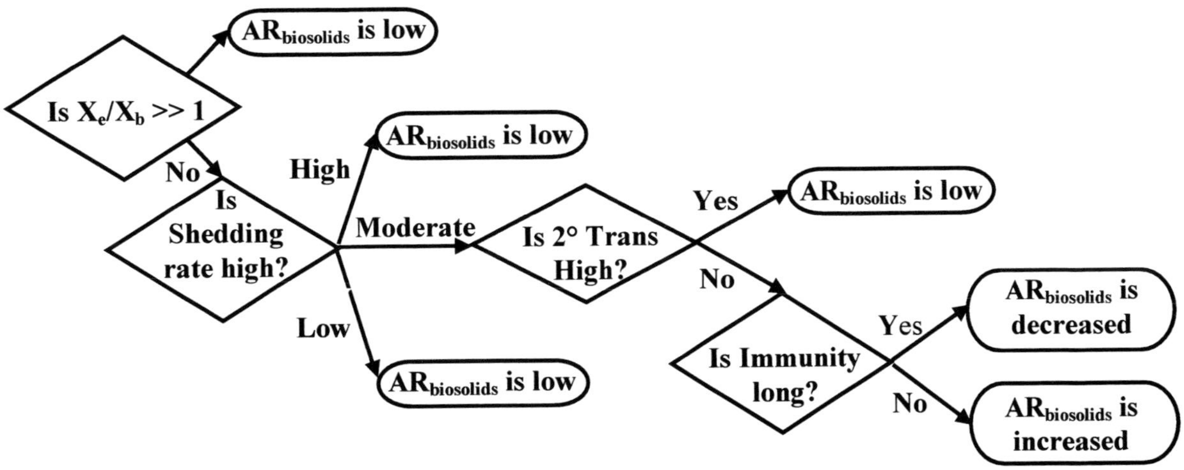

Figure 3-1. Decision tree for classifying the attributable risk associated with biosolids (AR_b) as high or low

3.2.1 Relative Contribution of Biosolids to Exposure

The most important parameters affecting the attributable risk associated with biosolids, AR_b, were X_b, the attenuation of pathogens in the biosolids, and X_e, the attenuation of pathogens in the environment other than biosolids. Both parameters X_b and X_e range from 0 to 1, where 0 reflects complete attenuation and 1 reflects no attenuation. Increasing the attenuation parameter, therefore, decreases the level of attenuation. Alternatively, these parameters also represent the contribution of either biosolids or the environment to the pathogen load. In this perspective, a large value (closer to 1) reflects a large contribution and a small value represents a small contribution.

The project team found that if X_e was large compared to X_b, then AR_b was low. In this case, biosolids treatment was more efficient at eliminating pathogens than other treatment processes such as those associated with water or food. As a result, most pathogens at exposure sites came from sources other than biosolids. Conversely, if X_b was large compared to X_e, then just the opposite was true (i.e., biosolids treatment was less efficient), and AR_b had the potential to be high depending on other parameter values.

Based on the simulations, the average AR_b for all simulations where X_e was set to a medium or high value was only 0.001 cases/person/year. When X_e was set to a low value and X_b was set to a medium or low value, the average AR_b for all simulations increased about tenfold to 0.013 cases/person/year. When X_e was low and X_b was high, the average AR_b increased tenfold yet again to 0.134 cases/person/year (refer to CART output in Appendix F). Clearly, more work is needed to provide rigorous estimates of these parameter values; however, these simulations illustrate the basic property that both the relative amount of pathogens diverted to biosolids versus other environmental exposure sites, and the relative attenuation of the different environmental processes compared with biosolids can be important determinants of risk.

To provide a better understanding of these results, the process of obtaining these risk estimates can be compared to a more traditional approach to estimating risk. In a more standard risk assessment, based on the chemical risk paradigm, AR_b would be estimated by collecting data at the exposure site to estimate dose, and, in turn, using this dose estimate as input to a dose-response function to estimate risk. This approach provides an estimate of the attributable risk associated with biosolids exposure. The basic assumption is that there is no interaction between biosolids exposure, human shedding of pathogens, person-to-person transmission, and environmental (other than biosolids) exposure. Given no interaction, this risk estimate is simply the first term of the attributable risk difference used in this approach.

Biosolids attributable risk is the difference in risk when all exposure pathways are present minus the risk when the biosolids pathway is eliminated. The traditional approach assumes that the second term (the risks when the biosolids pathway is eliminated) is zero. That is, the risk attributable to biosolids is simply the direct risks coming from exposure to biosolids. By incorporating these interactions, as demonstrated in the transmission model presented for this study, the risk estimates associated with biosolids exposure is achieved by looking at the difference between the risk when biosolids are present and when biosolids are absent. In this way risk becomes a function of an additional set of factors not accounted for in the chemical risk paradigm.

In the scenario described here, low, medium, and high values of X_b were set to 0.0001, 0.01, and 0.1, respectively. It was found that anytime X_b was set to 0.0001, AR_b was approximately zero regardless of other parameters; however, when X_b was 0.01, conditions existed in which AR_b could become significant (e.g. when X_e was low). Thus, in this scenario reducing X_b to 0.0001 is desirable because it virtually eliminates disease attributable to biosolids. In this model it was assumed that the attenuation of pathogens within biosolids is estimated by the product of each biosolids process being modeled. The log transformation of the parameter X_b provides an estimate of treatment efficiency in units of log removal. The log removal, therefore, is computed by the following equation:

$$\ln(X_b) = \ln(r \cdot s \cdot t_s \cdot a \cdot l \cdot t_l) \tag{3.2.1-1}$$

where r = treatment reduction, s = storage reduction rate, t_s = storage time, a = application reduction, l = die-off rate, and t_l = time to exposure; a practical way to reduce X_b would be to increase storage time.

To illustrate how the model can be used to assess a risk management decision, suppose that the analysis of the model suggested that reducing X_b to 0.0001 was desirable given independent data on estimates of X_e. And suppose that, based on estimates of the parameters, r, s, t_s, a, l, t_l, an additional removal of 0.7 log was necessary to achieve $X_b = 0.0001$. One management option would be to examine increasing the storage time, t_s. Doubling the storage time, for example, from 90 days to 180 days would provide the additional 0.7 log removal sufficient to accomplish this reduction.

One caveat to examining the ratio of X_e to X_b is that the results presented here are based on the assumption that f, the fraction of environmental pathogens associated with biosolids, is known. In general, the expression for the relative contribution of biosolids to environmental exposure to pathogens should include the ratio $f/(1-f)$, as shown above. Alternatively, the interpretation of X_b and X_e could include the factor f.

3.2.2 Shedding

The only known source of enterovirus pathogens is from human hosts. The mechanism from which pathogens are present in the environment, therefore, is the shedding of the virus by infectious individuals. One potential pathway consists of pathogens entering the wastewater stream and being transported into biosolids. In this model, the shedding process is represented by the parameter ϕ, the shedding rate of pathogens from infectious individuals into the environment. The parameter, ϕ, is defined as pathogens shed per infectious individual per day. Analysis of the simulations suggest that the rate in which pathogens are shed becomes important if $X_b >> X_e$. That is, shedding plays a role in biosolids associated risk only if pathogens are present in biosolids in sufficient quantity relative to other environmental sources. Under this condition, therefore, shedding may play a role in the level of risk associated with biosolids exposure.

Specifically, the analysis suggested that the attributable risk due to biosolids exposure, AR_b, may be large if the shedding rate is a moderate value (in this model a "moderate value" was somewhere between 10^5 and 10^7 pathogens/person/day). To understand this relationship better one can examine the effects of shedding on both the incidence of disease and the risk attributable to biosolids exposure. The assumption of the case study was that all infections ultimately came from other infectious individuals; i.e., pathogens did not come from outside of the community. Based on this assumption, low rates of shedding resulted in an estimate for incidence close to or at zero (in this model a "low value" was less than 10^5 pathogens/person/day). That is, there were not enough pathogens shed into the environment to sustain transmission via environmental exposure.

Another way to view this result is to observe that the pathogens released into the environment are attenuated through dilution, natural die-off, or treatment. Only a fraction, therefore, make it to an exposure site in which an individual may ingest a pathogen and become infected. If the number of pathogens released by infectious individuals during its infectious period results, on average, in less than one case of infection through environmental exposure, then the disease will not persist through this environmental loop alone. For example, in this simulation, if 100 infectious individuals shed pathogens into the environment at a rate of 10^4 pathogens/person/day, then these pathogens will potentially result in further infection. If, however, all the pathogens shed into the environment by these 100 infectious individuals during their duration of infectiousness produce fewer than 100 infections, then the pathogen cannot persist through this environmental pathway. Given this zero estimate for incidence, the attributable risk was also zero.

When the shedding rate was set to a high value, the simulation results suggested that the environmental exposure was also high, resulting in a large incidence estimate. The incidence estimate of 3.5 cases per person per year (Figure 3-2a) may be unrealistically high, suggesting that either the shedding rate estimate was too high or attenuation estimates were too low. For the purposes of this report, however, the concern was to demonstrate the potential relationships between model parameters and risk and not produce a precise estimate of risk. This high rate of shedding resulted in a high rate of exposure from all environmental exposures. Any additional contribution from biosolids was insignificant. Therefore, AR_b was close to zero for high values of ϕ (Figure 3-2b).

Figure 3-2a shows disease incidence when biosolids were present and when biosolids were absent. As just observed, the difference between these two plots (AR_b) was negligible at high values and low values of ϕ. However, at moderate values of ϕ there was a noticeable difference. For example, when $\phi = 10^4$ the number of cases per person per year was the same (zero) regardless of biosolids presence. Similarly, when $\phi = 10^8$, the number of cases per person per year was approximately 3.5 with or without biosolids. However, for $\phi = 10^6$ the number of cases per person per year was 2.3 when biosolids were present compared to 2.0 when biosolids were absent. The difference between the two scenarios is plotted in Figure 3-2b. The difference reached a maximum at approximately 10^5 and decreased for higher values of ϕ. The analysis revealed that, given $X_b >> X_e$, the average AR_b for all scenarios where ϕ was set to a medium value was 0.33 cases/person/year. When ϕ was set to a high or a low value, AR_b was only 0.036.

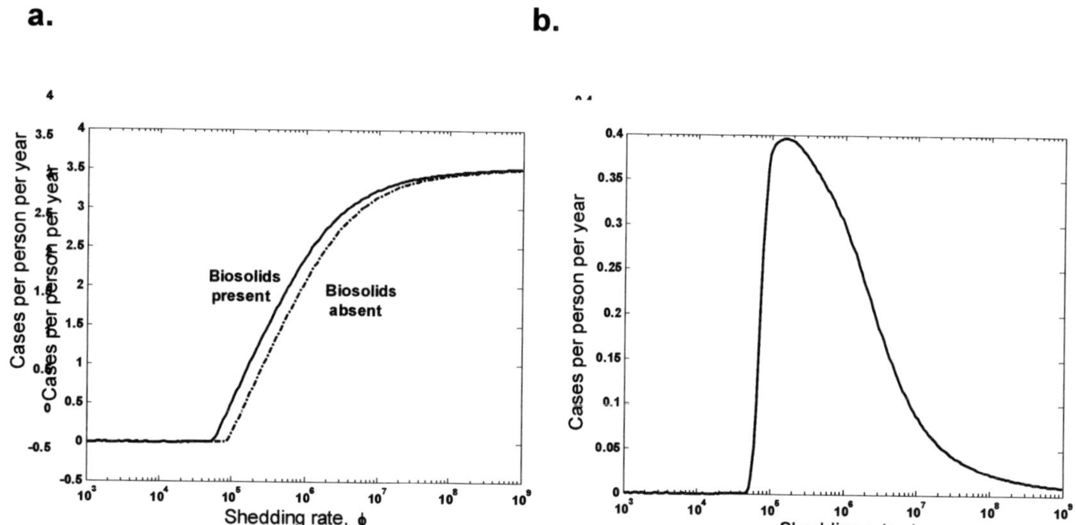

Figure 3-2. (a) Disease Incidence as a Function of ϕ, Shedding Rate, When Biosolids Were Present and When Biosolids Were Absent; (b) Attributable Risk (Excess Risk) Associated With Biosolids Exposure (the Difference in the Two Incidence Curves Shown in a). Plots are based on the condition that the ratio X_e/X_b is not >> 1. All other parameters, besides ϕ, were set to medium values.

The suggestion that shedding rates may affect risk estimates of exposure to biosolids emphasizes the fact that enteroviruses present in biosolids originate from human sources. These pathogens are transported through the environment from contamination site to exposures site. There exist many potential control points along this environmental pathway that can help to attenuate the concentration of pathogens and therefore decrease the risks associated with exposures to biosolids material. The introduction of shedding into the risk assessment highlights how the different components of an infectious disease process interrelate in estimating risk.

3.2.3 Secondary Transmission

Secondary transmission was defined as the transmission of a pathogen from an infectious person to a susceptible person. This may occur within communal settings such as a household or day care center, in which the transmission event occurs through an intermediary fomite such as a sink or toilet. Figure 3-3a suggests that as secondary transmission increases the incidence of disease within the community increases. Therefore, secondary transmission can serve to amplify an index case that is obtained from an environmental source, both those associated with biosolids and those associated with other sources. The two curves shown in Figure 3-3b were compared to assess the importance of secondary transmission to risks associated with biosolids. The difference between these two curves, shown in Figure 3-3b, provides this estimate of the attributable risk associated with biosolids exposure for different levels of secondary transmission.

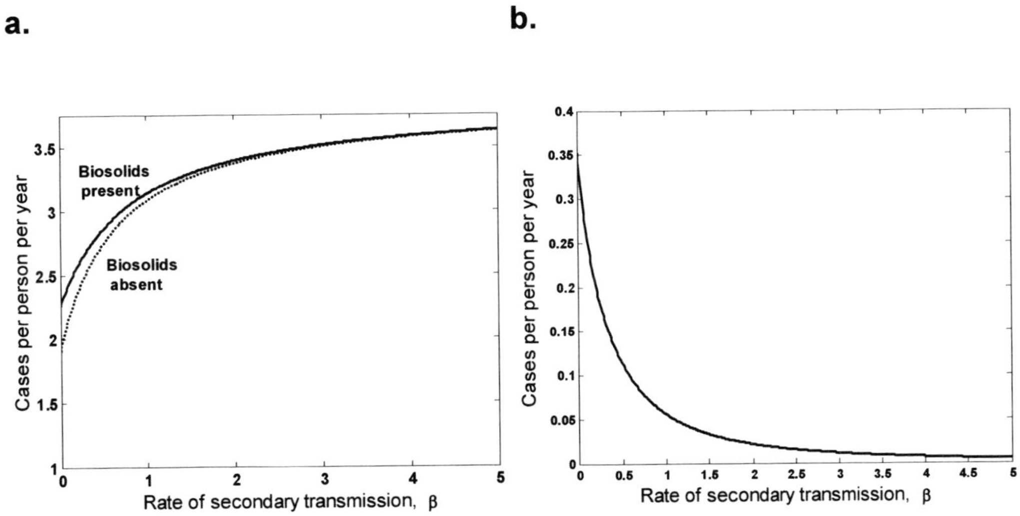

Figure 3-3.(a) Disease Incidence as a Function of β_{s2}, the Rate of Secondary Transmission, When Biosolids Were Present and When Biosolids Were Absent;(b) Attributable Risk (Excess Risk) Associated with Biosolids Exposure (the Difference in the Two Incidence Curves Shown in (a). Plots are based on the conditions that the ratio X_e/X_b is not >> 1 and that the shedding rate is "moderate" (see text for details). All other parameters, besides the transmission rate β_{s2} are set to medium values.

Again, estimating the biosolids associated risk in the chemical risk paradigm is accomplished by simply measuring the concentration of biosolids at the exposure site. This assumes that other transmission pathways, such as secondary transmission, are zero. Once these other transmission pathways are accounted for, the appropriate risk estimate becomes the difference in incidence between the condition in which there is no biosolids exposure and the condition in which there is biosolids exposure.

As summarized in Figure 3-1, secondary transmission was not an important determinant of biosolids associated risk unless the relative contribution of biosolids versus other environmental sources was high ($X_b >> X_e$), and the shedding rate, ϕ, was "moderate" (between 10^5 and 10^7 pathogens/person/day). Under these conditions, increased person-to-person transmission serves to dilute the impact of other competing transmission pathways and drives the value of AR_b down; that is, when person-to-person transmission is high, the likelihood of developing disease from an environmental pathway is small. Figure 3-3b shows the difference between these two scenarios (AR_b). The difference is largest when secondary transmission is zero. At this point all disease is attributable to environmental exposure sources. As β_{s2} increases the two plots converge and the difference approaches zero. Note that when β_{s2} is large ($\beta_{s2} > 4$), AR_b is small (~ 0), however, the incidence of disease in the community is large (~ 3.5).

The finding that secondary transmission serves to decrease the attributable risk associated with biosolids is dependent on the fact that we are examining a single community endemic condition in which biosolids are one of many environmental sources of a given pathogen. There may be other conditions, such as the case in which biosolids serves as a potential risk to a point source outbreak of a unique viral strain that has no other environmental sources. Under these conditions, secondary transmission may serve to amplify not only the overall incidence of disease within the community, but also the risk associated with biosolids exposure.

3.2.4 Immunity

After an individual recovers from an infectious period, this model allows for that individual to be partially and temporarily protected from further infection. The average time of protection is measured by the inverse of the parameter γ ($1/\gamma$). During this period, $1/\gamma$, the level of protection is assumed to linearly decrease from full protection to no protection. This conferred immunity serves to limit the number of individuals at risk at any given time. The smaller the duration of protection, the greater the number of infections possible in a given year. This is illustrated in Figure 3-4a where the cases per person per year decrease with increasing duration of protection.

Conversely, as the duration of protection increases, a larger proportion of the population is in the protected state and a lower proportion is completely susceptible. Under these conditions, the overall incidence in the community is low. If the overall risk of disease for a particular pathogen is low then the attributable risk associated with biosolids will also be low. Long durations of immunity (small γ values), therefore, lowered the attributable risk associated with biosolids, AR_b. As can be seen in Figure 3-4b, the difference in disease incidence when biosolids were present vs. when biosolids were absent is greatest for short durations of immunity. This difference decreases as duration of immunity increases.

The model structure used in this study assumed that enteroviruses were a homogeneous group of viruses in which exposure to one strain confers protection from infection of any other strain. Although, some degree of cross-immunity between strains is plausible, it is likely that there is not complete cross immunity across all strains. Further work is needed to examine the effects of different levels of cross-immunity.

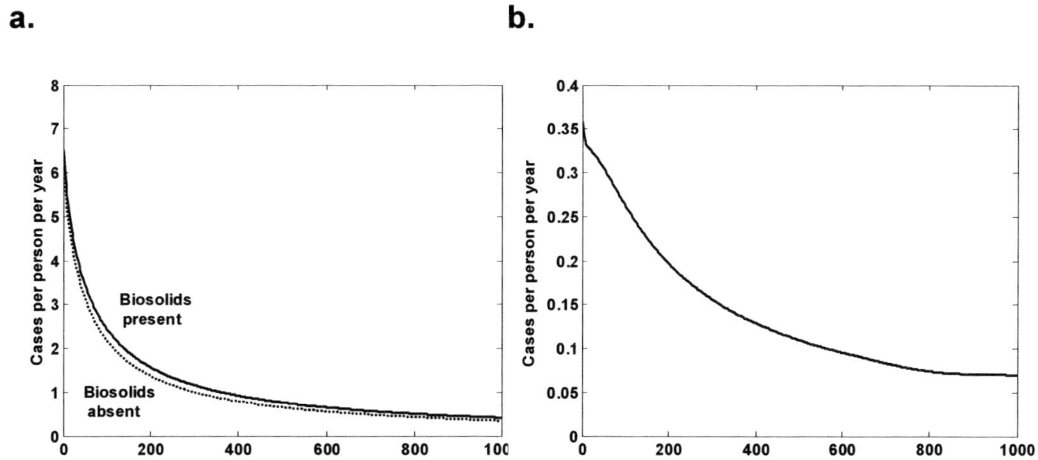

Figure 3-4. (a) Disease Incidence as a Function of $1/\gamma$, the Duration of Immunity, When Biosolids Were Present and When Biosolids Were Absent; (b) Attributable Risk (Excess Risk) Associated with Biosolids Exposure (the Difference in the Two Incidence Curves Shown in a). Plots are based on the condition that the ratio X_e/X_b is not >> 1, the shedding rate is "moderate" (see text for details), secondary transmission is negligible. All other parameters are set to medium values.

3.2.5 A Decision Tree for Estimating Risks

The first decision node of the decision tree for determining the level of risk associated with biosolids exposure (see Figure 3-1) addresses the environmental exposure component of the risk model. It is at this level that risk managers can examine the effects of treatment and processes options to ascertain the costs of ensuring a low attributable risk. Different processing options, therefore, can be evaluated to determine whether or not a given change in the process will have an effect on the first node decision. The decision at this node is whether or not the contribution of biosolids exposure is significant compared with the contribution of other sources. The actual determination of when, for example, X_e/X_b is not >> 1 is a risk management decision that is dependent on what level to set tolerable risk. An example examining the increase of storage time was discussed in Section 3.2.1.

If the decision at the first node in Figure 3-1 is no, then the shedding rate of the pathogen into the environment, secondary transmission rates, and the duration of protection become the important factors that determine whether or not the attributable risk associated with biosolids is predicted to be low.

Decision trees like that shown in Figure 3-1 can be constructed in a risk assessment and can provide the risk manager and regulator with information on the source of the risks and benefits associated with different control options. For example, various biosolids process-specific parameters, such as storage time or dilution of soil, can be varied to impact, in turn, the initial node decision of Figure 3-1. Wastewater treatment can be increased to counteract high shedding rates and impact the second decision node. Finally, educational programs for parents or those in charge of communal settings, such as daycare centers, can be implemented and impact the third decision node. In addition to estimates on risks and benefits, specific sensitivity information on the decision to use a given control option can be obtained by examining the effects of varying different parameters in the model. For example, simulations can be conducted to provide information on impact of changing the storage time for different levels of shedding and secondary transmission.

3.3 Conclusions

In 1993, EPA was unable to base pathogen regulations for the Part 503 rule on risk assessment because the necessary methodologies were not sufficiently developed. Since that time, advances in microbial risk methodology have provided the means to conduct these risk assessments. This project has combined and adapted these advances to address human health risks from the exposure to pathogens within biosolids. With additional data collection and adaptation of appropriate exposure modeling methods, this approach has potential to be applied at both the regulatory and operational levels. At the regulatory level, this tool can be useful as part of a process to review and revise the pathogen regulations for the Part 503 rule. At the operational level, this tool can be useful in quantifying differences in potential health impacts, or as part of the evaluation of environmental conditions and biosolids treatment processes.

One particular advance in the area of microbial risk assessment is the development of health risk models that explicitly account for the different transmission pathways that can result in human infection. The inclusion of these transmission pathways provided insight that would not be predicted by a chemical-based model. For example, one conclusion from the simulations was that intermediate shedding would result in higher attributable risk than high rates of shedding. This is because high rates of shedding results in high levels of pathogens in the environment, which, in turn, can result in pathogens circulating in the community at a high rate. Infection, therefore, spread throughout the community from person-to-person or person-to-environment-to-person. Therefore, eliminating any particular environmental pathway will have little effect on the overall level of disease. By contrast, the chemical model would simply conclude that higher levels of shedding result in high levels of pathogens and, therefore, higher levels of risk.

These results were presented in this report through a structured decision tree of high and low attributable risk estimates for different environmental conditions. Describing the model output in the form of a decision tree can be a useful tool for the regulator by allowing the identification of factors that are the most important determinants of high and low risk conditions. In this form, the decision tree provides an assessment of the sensitivity of parameter values to risk levels. Although not a formal part of this project, an uncertainty analysis could inform the regulator as to what additional research and data are needed to improve a given regulatory decision. This same decision tree can also be a useful tool at the operational level for outlining the health benefits associated with improving different components of the biosolids treatment process, for example, to determine the pathogen attenuation required to attain a low-risk condition.

This study showed that a disease transmission model can provide insight into risk estimates that cannot be attained in a chemical risk model. Another significant result is that risk assessment models can also provide information on data needs. These data needs are crucial in the ability to estimate illness rate within a community due to a biosolids land application program or establish the adequacy of existing standards. Surveillance data for pathogens from various points in the environmental transport process from wastewater to biosolids provides information that can be used to improve risk estimates. Other types of environmental data useful when analyzing different scenarios are measurements of die-off both from natural and treatment factors. It is additionally valuable to have data that provides information on the functional relationship between these die-off factors and various environmental factors such as temperature and humidity. The health effects data required are available to some degree in the literature. More detailed data on household-level transmission and the immune process would provide the needed information to incorporate additional structure into the model. In particular, a better understanding of how the different strains of enteroviruses interact in developing immunity would be immensely useful.

To demonstrate the potential application of this approach a more detailed and realistic risk assessment is needed. This risk assessment activity would require a comprehensive definition of relevant biosolids application scenarios and definition of relevant risk measures. This will provide the bridge between the model development and demonstration discussed in this report, and an actual risk assessment. Scenarios that will be the most useful will be those that are framed as a comparative analysis. For example, it would be interesting to compare occupational exposures, which are generally associated with higher levels of exposure, longer exposure times, and fewer individuals exposed, with consumer exposures, which are associated mostly with lower levels of exposure, shorter exposure times, and larger number of individuals exposed. Every region in the country will have different exposure scenarios that are relevant to the particular conditions in which the biosolids are produced and used. For example, some may be more concerned with children exposed to biosolids applied near housing tracts whereas others may be more concerned with the potential of water runoff to contaminate sources of recreational or drinking water.

Given the increasing interest and ongoing controversy in the beneficial uses of biosolids and the rapid growth in the public use of those biosolids, this is the time to develop microbial risk assessment techniques for application to biosolids regulatory and operational decision making.

APPENDIX A

CONCEPTUAL MODELS FOR EXPOSURE TO PATHOGENIC ORGANISMS

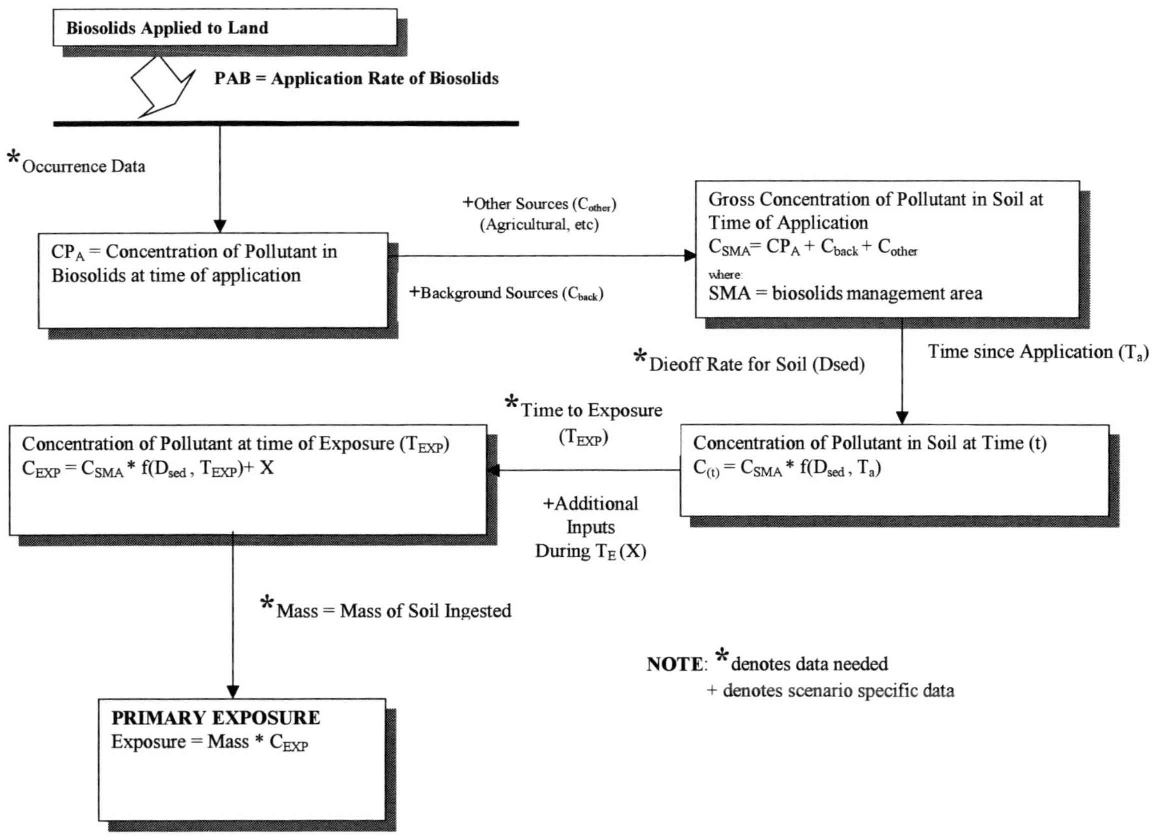

Figure A-1. Conceptual Model for Exposure through Direct Ingestion of Soil

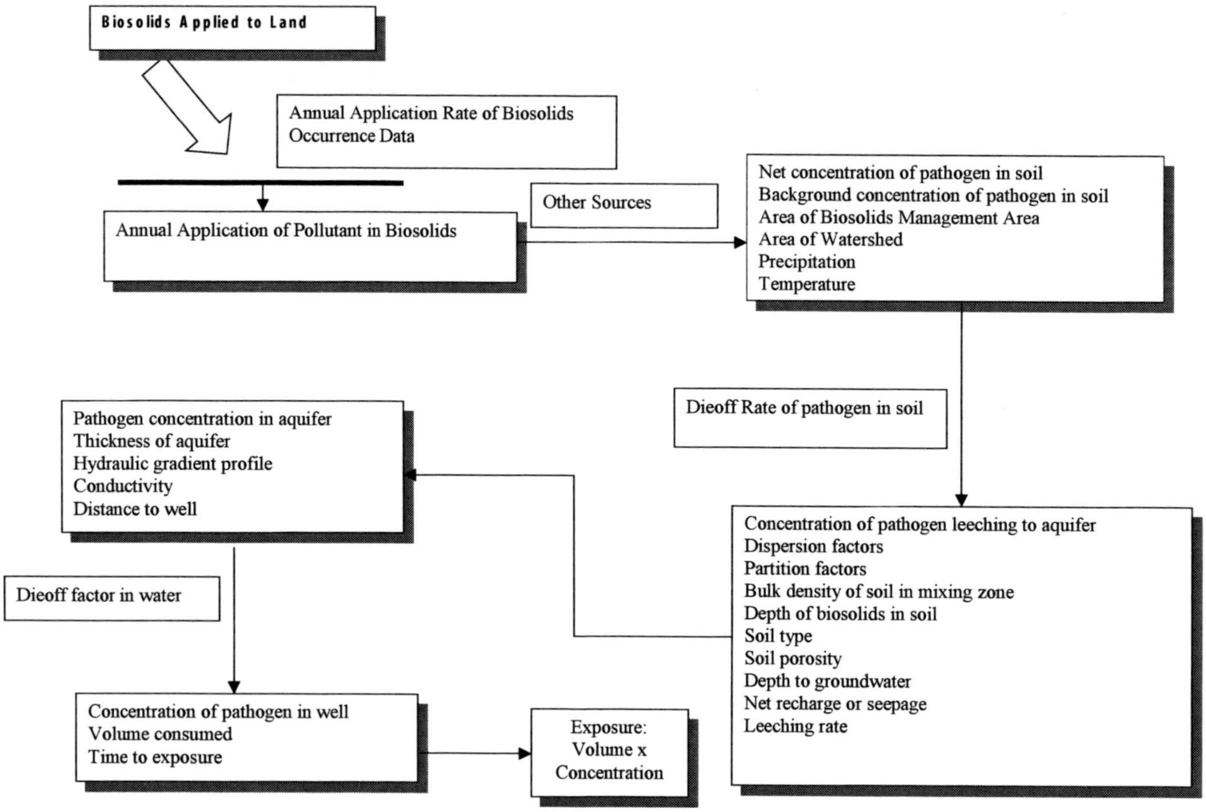

Figure A-2. Conceptual Model for Exposure through Ingestion of Ground Water

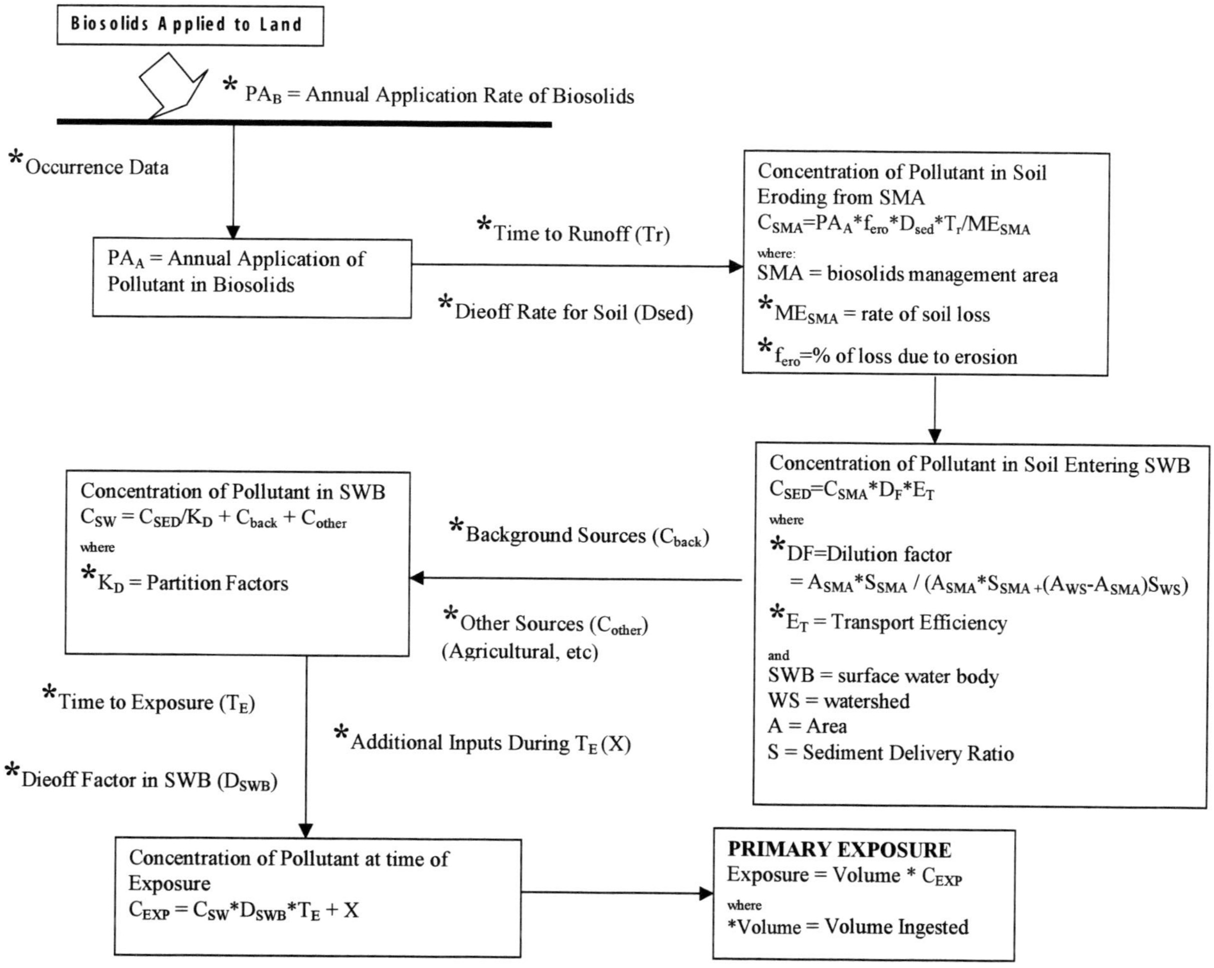

Figure A-3. Conceptual Model for Exposure through Ingestion of Surface Water

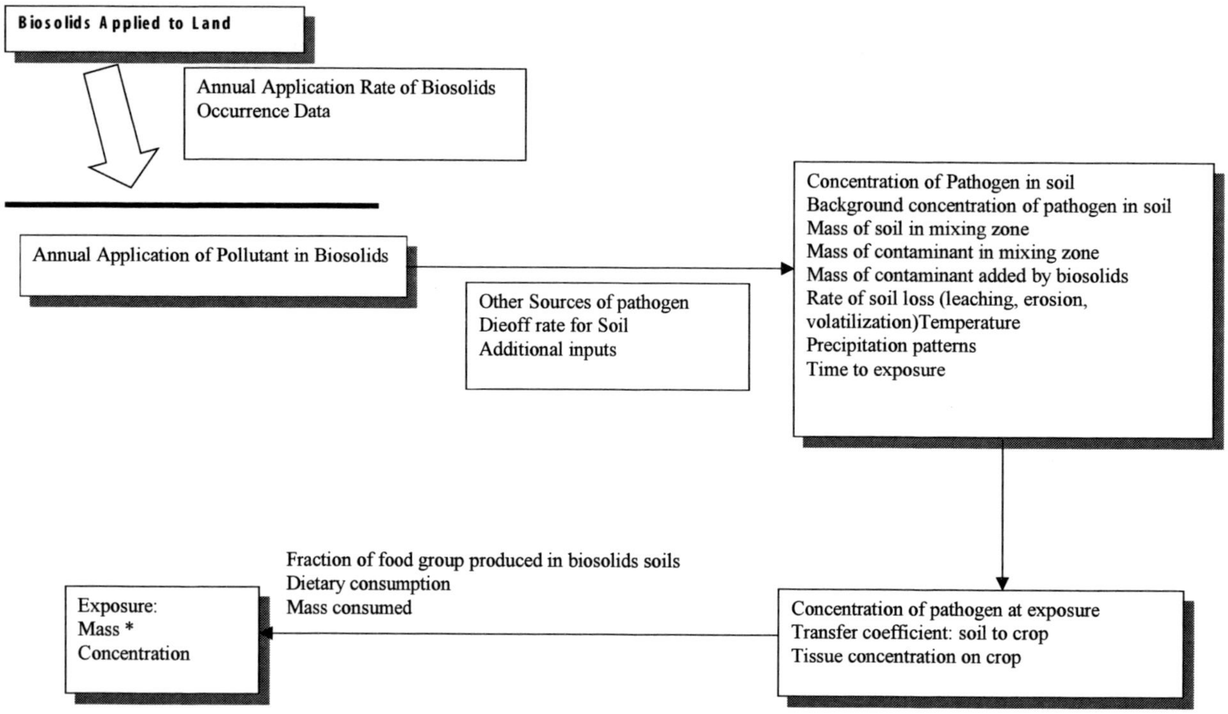

Figure A-4. Conceptual Model for Exposure through Plant Consumption

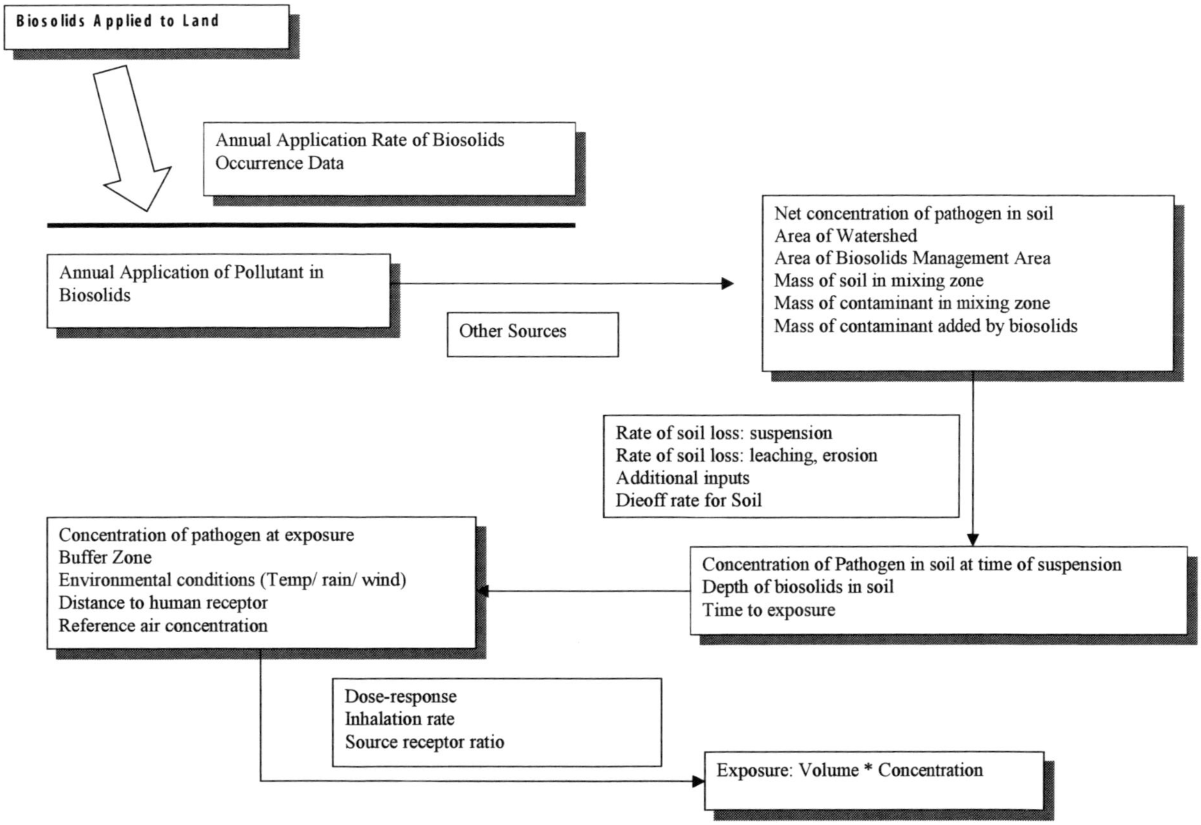

Figure A-5. Conceptual Model for Exposure through Inhalation

APPENDIX B

EXPOSURE PROFILE (ENTEROVIRUSES)

Enterovirus occurrence in anaerobically digested biosolids. Monpoeho et al. (2000) report that a quantitative reverse transcriptase polymerase chain reaction (RT-PCR) method is now available that can be used to determine the amount of enteroviruses in urban sludge samples. Unfortunately, no data have yet been reported in the literature using that method.

In a study that was described previously, Tata and co-workers (2000a) provide an estimate of 0.117 plaque-forming units (PFU)/4g for enteric virus density in anaerobically digested sludge (20 days at 35°C). In a study carried out in 1995 by Gibbs and co-workers (1995) enteroviruses were not detected in sludge that had been treated by mesophilic anaerobic digestion and mechanical dewatering. Goddard and co-workers (1981) report enterovirus concentrations ranging between 5–200 PFU/g in mesophilic digested sludge. Other specific data quantifying enterovirus concentrations in biosolids generated via mesophilic anaerobic digestion were not found in the literature.

Enterovirus occurrence in raw and secondary sludge. The levels of enteroviruses reported in the literature for raw and secondary sludge are relatively consistent from investigation to investigation, although the concentration ranges reported are highly variable.

Mignotte and co-workers (1999) report densities of infectious enteroviruses in activated sludge ranging from <3 to 300 most probable number (MPN)/10 g dry mass, and from 10 to 2200 MPN/10g dry mass in raw sludge. For 10 g dry mass of primary sludge, Hu and co-workers (1997) detected 16,000 infectious units, Soares and co-workers (1994) 330 MPN, and Pederson (1981) 3900 PFU. Similarly, Goddard (1981) reported 30–900 PFU/g. Albert and Schwartzbrod (1991) reported 30–135 PFU/100 mL, which is consistent, albeit on the low end of the range of 10^2–10^4/ mL reported in the comprehensive review assembled by Sorber and Moore (1986). At the low end of the densities reported are the results from the recent investigation reported by Tata and co-workers (2000a) in which they report 3.6 to 3.83 PFU/4g total solids dry weight in their raw sludge.

Reduction of enteroviruses during mesophilic anaerobic digestion. The reduction of enteroviruses during mesophilic anaerobic digestion is well characterized by data in the literature. The project team estimated that mesophilic anaerobic (or aerobic) digestion will reduce viruses by 0.5 to 2.0 logs, based on investigations or reviews carried out by Tata and co-workers (2000a), Ahmed and Sorensen (1997), Sorber and Moore (1986), and Feachem (1983).

Storage of Biosolids. Based on laboratory analysis, Ahmed and Sorensen (1995) reported the rate of destruction of stored pathogens in biosolids. In that study, mean decay rates for poliovirus are given for temperature ranging from 5°C to 50°C. At 20°C, the mean decay rates in log reduction/day (for seeded biosolids) was estimated at 0.084 for poliovirus. Given that autoheating did not occur during the follow-up study (1997) for storage piles of biosolids, 20°C would be consistent with ambient conditions for the case study. (Note that the decay rate for 5°C was 0.02 log/day.)

Survival in soil. The most applicable data for this investigation come from a review of the literature on the survival and transport of pathogens in sludge amended soil carried out by Sorber and Moore (1986).

Based on investigations summarized in Sorber and Moore (1986) from Bitton (1984), Tierney (1977), Hurst (1989), and Farrah (1981), the project team estimated a t_{90} of 1–4 days for enteroviruses. It should be noted, however, that survival is much longer at cooler (e.g., winter in Cincinnati t_{90} = 30 days) rather than warmer temperatures (18 °C–27°C).

Reddy and co-workers (1981) derived an average die-off rate constant (d^{-1}) of 1.45 for viruses in a soil-water-plant system (T_{90} equivalent =3.3 day). Hurst (1989) developed a regression equation (based on data from Hurst (Akin and J.C. 1978) and Farrah (1981)) describing inactivation of indigenous viruses following land surface spreading of sludge. The reduction was estimated to occur linearly as follows: 0.266±0.031 (where reduction is log/d).

Soil ingestion. Ingestion of soil by humans has been a documented phenomenon for centuries and still takes place today according to a review of the literature (Simon 1998). EPA has assumed that 95% of children ingest 200 mg soil/day or less for exposure assessment purposes (USEPA 1989b). Although some children may ingest large amounts of soil per day (25–60 g/day) (Calabrese et al. 1997a), most of the reported literature is in relative agreement with EPA rates for children noted above.

In an investigation estimating soil ingestion for children at a Superfund site, Calabrese and co-workers (1997b) estimate median levels of ingestion to be < 40 mg/day and 95th percentiles to be > 160 mg/day. Stanek and Calabrese (1995) analyzed data from a 1989 soil ingestion study in Amherst, Massachusetts, and report a median daily soil ingestion of 13 mg/day or less, and a 95th percentile of 138 mg/day or less. Mean soil ingestion estimates for an 8-day period were 45 mg/d or less for 50% of the children, whereas 95% of the children reported a mean soil ingestion of 208 mg/day or less.

In 1991, Calabrese and Stanek reviewed the previously carried out studies to qualitatively and quantitatively assess the extent of soil ingestion in children using tracer studies (Calabrese and Stanek 1991). They conclude that Binder et al. (1986) and van Wijnen et al. (1990) "provide no convincing evidence to support qualitative and quantitative estimates of soil ingestion due to inherent limitations in their study design." LaGoy (1987) derived estimates of average soil ingestion rates of 25 to 100 mg/day, based on data by Binder et al and van Wijnen et al.

Clausing and co-workers (1987) measured tracers in the feces of children and in soil and estimated a soil ingestion rate of 50 to 100 mg/day. Earlier estimates including those by Hawley (1985), and Lepow and co-workers (1975, 1974) are reasonably consistent with those summarized above.

APPENDIX C

MATHEMATICAL EQUATIONS FOR DISEASE TRANSMISSION MODEL

This appendix presents the equations that correspond to the model depicted in Figure 2-8. The model is defined by the following seven state variables:

S	Number of Individuals in Susceptible State
E	Number of Individuals who have been exposed but are non-infectious
C_1	Number of Individuals in Carrier State (asymptomatic & infectious)
D	Number of Individuals in Diseased State (symptomatic & infectious)
C_2	Number of Individuals in Carrier State (asymptomatic & infectious)
P	Number of Individuals in Post-Infection State
W	Number of organisms at the exposure site

and the following parameters:

α		Transition rate out of exposed state. Inverse of incubation period.
P_{sym}		Probability of symptomatic response
δ		Transition rate out of diseased state. Inverse of disease duration
σ_1, σ_2		Transition rate out of carrier state. Inverse of duration of shedding.
γ		Transition rate out of post infection state. Inverse of duration of immunity
β	β_{s1}	Transmission rate (susceptible to exposed from environmental exposure)
	β_{s2}	Transmission rate (susceptible to exposed from person-to-person contact)
	β_{p1j}	Transmission rate (Post infection state *j* to exposed from environmental exposure)
	β_{p2j}	Transmission rate (Post infection state *j* to exposed from person-to-person contact)
ϕ		Rate that infectious individuals shed into the environment
μ		Pathogen die-off rate

Movement within the **E**, **C_1**, **C_2**, **D**, and **P** states is characterized by a distributed delay. This delay is realized by a series of ordinary differential equations. In such a characterization, the time spent in any state is described by a gamma distribution defined by two parameters, *n*, the number of first order processes, and *k*, the rate constant governing the movement of one first-order process to the next. For example, the following set of equations depict the transition through the exposed state, where I_i is the rate of entering and I_o is the rate of leaving the exposed state, and *k* is equal to α.

$$\frac{dE_1}{dt} = I_i - \alpha E_1$$

$$\frac{dE_2}{dt} = \alpha E_1 - \alpha E_2$$

$$\frac{dE_3}{dt} = \alpha E_2 - \alpha E_3$$

$$\frac{dE_4}{dt} = \alpha E_3 - I_o$$

The parameters *n* and *k* determine the mean and variance of the gamma distribution:

$$Mean = {}^{n}\!/_{k}$$

$$Varience = {}^{n}\!/_{k^2}$$

For computational purposes, n was set to four in all cases. The transition rate k was replaced with a corresponding transition parameter for the given state (e.g. α for the transition through the exposed state). The mean transit time for each state, therefore, is 4 divided by the corresponding transition rate parameter.

The change in the number of susceptible individuals is governed by three possible events. Two are transmissions events that result in a decrease in the number of susceptible individuals through either contact with an infectious individual (a individual in either the **D**, $\mathbf{C_1}$, or $\mathbf{C_2}$ state), or contact with a contaminated environment, $\mathbf{W_4}$. The third is a transition event in which a protected individual becomes susceptible again. The following equation describes this change in the number of susceptible individuals.

$$\frac{dS}{dt} = -\beta_{s2}\, S(D + C_1 + C_2) - \beta_{s1}\, S\, W_4 + \gamma P_4$$

For detail of the parameters and their values see the text in Chapter 2.

To incorporate the concepts of incomplete and waning immunity, four protected states ($\mathbf{P_1}$, $\mathbf{P_2}$, $\mathbf{P_3}$, $\mathbf{P_4}$) were included in the model. In the first state ($\mathbf{P_1}$), all individuals are completely immune to re-infection. In the subsequent three states ($\mathbf{P_2}$, $\mathbf{P_3}$, $\mathbf{P_4}$) reinfection becomes possible to a varying degree; in each successive protected state the level of protection decreases until, finally, the population is completely susceptible once again. As a result, movement from three of the protected states ($\mathbf{P_2}$, $\mathbf{P_3}$, $\mathbf{P_4}$) to the exposed state (**E**) is possible and is regulated by the following relationship:

$$\beta_{pji} = \frac{\beta_{sj} \cdot (i-1)}{n}$$

where j takes on the value of 1 or 2 corresponding to environmental or person-to-person transmission; i takes on the value of 1, 2, 3, or 4 corresponding to the particular protected state; and n corresponds to the total number of protected states (four in this case). The following equation describes the immune process.

$$\frac{dP_1}{dt} = \sigma_1\, C_{14} + k_{\sigma 2}\, C_{24} - \gamma\, P_1$$

$$\frac{dP_2}{dt} = \gamma\, P_1 - \gamma\, P_2 - \beta_{p1_2}\, P_2\, (D + C_1 + C_2) - \beta_{p2_2}\, P_2\, W_4$$

$$\frac{dP_3}{dt} = \gamma\, P_2 - \gamma\, P_3 - \beta_{p2_3}\, P_3\, (D + C_1 + C_2) - \beta_{p2_3}\, P_3\, W_4$$

$$\frac{dP_4}{dt} = \gamma\, P_3 - \gamma\, P_4 - \beta_{p2_4}\, P_4\, (D + C_1 + C_2) - \beta_{p1_4}\, P_4\, W_4$$

Individuals enter the initial protected state, $\mathbf{P_1}$, from either of the two carrier states (see model diagram in Figure 2-1). The transition through the protected states occurs at a rate, k_γ. For the remaining protected states, $\mathbf{P_2}$, $\mathbf{P_3}$, $\mathbf{P_4}$, there are two transmission events possible on through contact with another infectious individual and the other through contact with a contaminated environment.

Once infected individuals are in the exposed state, that represents the state of being infected but not infectious. Four transmission events enter into the initial exposed state, $\mathbf{E_1}$. This state is governed by the following set of equations.

$$\frac{d\mathbf{E}_1}{dt} = \beta_{s2}\,\mathbf{S}(\mathbf{D}+\mathbf{C}_1+\mathbf{C}_2)+\beta_{s1}\,\mathbf{S}\,\mathbf{W}_4+\beta_{p2_2}\,\mathbf{P}_2(\mathbf{D}+\mathbf{C}_1+\mathbf{C}_2)+\beta_{p1_2}\,\mathbf{P}_2\,\mathbf{W}_4$$
$$+\ \beta_{p2_3}\,\mathbf{P}_3(\mathbf{D}+\mathbf{C}_1+\mathbf{C}_2)+\beta_{p1_3}\,\mathbf{P}_3\,\mathbf{W}_4+\beta_{p2_4}\,\mathbf{P}_4(\mathbf{D}+\mathbf{C}_1+\mathbf{C}_2)$$
$$+\beta_{p1_4}\,\mathbf{P}_4\,\mathbf{W}_4-\alpha\,\mathbf{E}_1$$

$$\frac{d\mathbf{E}_2}{dt} = \alpha\,\mathbf{E}_1-\alpha\,\mathbf{E}_2$$

$$\frac{d\mathbf{E}_3}{dt} = \alpha\,\mathbf{E}_2-\alpha\,\mathbf{E}_3$$

$$\frac{d\mathbf{E}_4}{dt} = \alpha\,\mathbf{E}_3-\alpha\,\mathbf{E}_4$$

Once an individual completes the transition through the exposed state, a proportion, p_{sym}, enter the diseased state, **D** (infectious and symptomatic) and then transition into a carrier state, $\mathbf{C_2}$. The remaining proportion, 1-p_{sym}, enters a carrier state, $\mathbf{C_1}$ (infectious and asymptomatic). The following equations describe the transition through the diseased and two carrier states.

$$\frac{d\mathbf{D}_1}{dt} = p_{sym}\,\alpha\,\mathbf{E}_4-\delta\,\mathbf{D}_1 \qquad \frac{d\mathbf{C}_{2_1}}{dt} = \delta\,\mathbf{D}_4-\sigma_2\,\mathbf{C}_{2_1}$$

$$\frac{d\mathbf{D}_2}{dt} = \delta\,\mathbf{D}_1-\delta\,\mathbf{D}_2 \qquad \frac{d\mathbf{C}_{2_2}}{dt} = \sigma_2\,\mathbf{C}_{2_1}-\sigma_2\,\mathbf{C}_{2_2}$$

$$\frac{d\mathbf{D}_3}{dt} = \delta\,\mathbf{D}_2-\delta\,\mathbf{D}_3 \qquad \frac{d\mathbf{C}_{2_3}}{dt} = \sigma_2\,\mathbf{C}_{2_2}-\sigma_2\,\mathbf{C}_{2_3}$$

$$\frac{d\mathbf{D}_4}{dt} = \delta\,\mathbf{D}_3-\delta\,\mathbf{D}_4 \qquad \frac{d\mathbf{C}_{2_4}}{dt} = \sigma_2\,\mathbf{C}_{2_3}-\sigma_2\,\mathbf{C}_{2_4}$$

$$\frac{d\mathbf{C}_{1_1}}{dt} = (1-p_{sym})\,\delta\,\mathbf{E}_4-\sigma_1\,\mathbf{C}_{1_1}$$

$$\frac{d\mathbf{C}_{1_2}}{dt} = \sigma_1\,\mathbf{C}_{1_1}-\sigma_1\,\mathbf{C}_{1_2}$$

$$\frac{d\mathbf{C}_{1_3}}{dt} = \sigma_1\,\mathbf{C}_{1_2}-\sigma_1\,\mathbf{C}_{1_3}$$

$$\frac{d\mathbf{C}_{1_4}}{dt} = \sigma_1\,\mathbf{C}_{1_3}-\sigma_1\,\mathbf{C}_{1_4}$$

The remaining set of equations describes the transition of the pathogens that are shed into the environment at a rate, ϕ, by the individuals in the three infectious state, **D**, $\mathbf{C_1}$, and $\mathbf{C_2}$. For this model, the transport process represents the movement of pathogens from the contamination to the exposure site. The pathogens move through the environment at a rate, k_w, and have a natural die-off rate, μ. This die-off parameter is meant to represent mortality factors not represented in terms that are part of the transmission rate parameter, β_{s1}, as described in Chapter 2. The following equations govern the pathogen transport process through the environment.

$$\frac{d\mathbf{W}_1}{dt} = \phi(\mathbf{D} + \mathbf{C}_1 + \mathbf{C}_2) - k_w \mathbf{W}_1 - \mu \mathbf{W}_1$$

$$\frac{d\mathbf{W}_2}{dt} = k_w \mathbf{W}_1 - \mu \mathbf{W}_2$$

$$\frac{d\mathbf{W}_3}{dt} = k_w \mathbf{W}_2 - \mu \mathbf{W}_3$$

$$\frac{d\mathbf{W}_4}{dt} = k_w \mathbf{W}_3 - \mu \mathbf{W}_4$$

APPENDIX D

HEALTH EFFECTS PROFILE (ENTEROVIRUSES)

The following information provides health effects information for enteroviruses needed to identify the model parameters. Members of the genus *Enterovirus*, a subgroup of the family Picornaviridae are small in size (27 nm) and contain ribonucleic (RNA) genomes. They all behave in a similar fashion biologically with regard to the human intestinal tract. Enteroviruses multiply throughout the alimentary tract and cause a variety of infections. Those that have undergone limited replication in the oropharynx survive transit through the stomach and implant in the lower intestinal tract. Following is a list of species and serotypes:

Genus *Enterovirus*	
Polioviruses of humans	3 (types 1–3)
Coxsackieviruses, group A	23 (types A1–A22, A24)
Coxsackieviruses, group B	6 (types B1–B6)
Echoviruses of humans	31 (types 1–9, 11–27, 29–33)
Enteroviruses of humans	4 (types 68–71)
Enteroviruses of other species	At least 34 types

Epidemiology of Enteroviruses. It has been reported that there are 20 million enteroviral infections per year in the United States (Pallansch 1997). Of 18,309 non-polio enteroviral isolates collected by the Centers for Disease Control and Prevention (CDC) between 1970 and 1979, echoviruses accounted for 58%, group B coxsackievirus for 24%, group A coxsackievirus for 8%, enterovirus 68–71 for 0.1%, and 10% of isolates were not classified (Moore 1982).

There were 3209 non-polio enterovirus (NPEV) cases reported to the CDC between 1993 and 1996. The predominant serotypes were echovirus 9 (12.7%), coxsackievirus B5 (11.5%), echovirus 30 (9.5%), coxsackievirus A9 (6.6%), coxsackievirus B2 (6.2%), echovirus 6 (5.1%), and echovirus 11 (4.5%). Echovirus 6, 7, 9, 11 and coxsackievirus A9, B2–4 accounted for 58.3% to 80.8% of isolates for each of the 4 years. All 10 of these were among the top 10 serotypes for the 1970–1983 surveillance period. However, echovirus 4, the fifth most prevalent between 1970–1983, was not among the top 15 isolates between 1993–1996 (CDC 1997).

Probability of Illness. In general, data in the literature suggest that approximately half of enteroviral infections are asymptomatic (Melnick 1996). For example, the New York Virus Watch (NYVW) program reported that illness was observed in 56% of confirmed echovirus infections and in 57.3% of confirmed coxsackievirus infections (Kogon et al. 1969). Probability of illness data for specific serotypes are summarized in Table D-1.

Immunity and Seroprevalence. In the NYVW program, it was observed that household transmission of coxsackievirus was lower among exposed family members who had detectable neutralizing antibodies at the beginning of the study (25% if antibodies present versus 76% if no antibodies). Neutralizing antibody to echovirus was highly protective since there was only one case of secondary infection in an immune family member. Furthermore, there was only one detected reinfection due to echovirus (serotype 25), whereas the overall rate of coxsackievirus reinfection was 25% (50% in children under 10, and 10% among teens and adults) (Kogon et al. 1969).

Table D-1. Probability of Illness Data for Specific Enterovirus Serotypes

Echo 4	During a 6-week outbreak of aseptic meningitis in Iowa, house-to-house surveillance found an overall attack rate of 191 cases per 1,000 population (range 46–308) (Moore 1982; Moore and Morens 1984).
Echo 9	Attack rates of echovirus 9 were 50-60% in children versus 17-33% in adults. (Moore 1982, Moore and Morens 1984)
Echo 30	Among the 64 families participating in the Seattle Virus Watch during an outbreak of aseptic meningitis, 47% of those infected showed signs of illness (33% for adults and 52% for children) (Hall et al. 1970). • During a 1997 outbreak of 21 cases of aseptic meningitis in rural Germany, a cross-sectional survey of 197 of the town's children under 16 years identified 31 (16%) children who met the case definition of enteroviral illness, but only 15 (48%) of these visited a doctor (Reintjes et al. 1999). • The attack rate of echovirus 30 per 1,000 in the U.K. was 19.7 for 0–9 year olds, followed by rates of 7.11, 4.85, 4.73, 1.5, and 0 for the succeeding 10-year cohorts (Moore 1982). • Attack rates for echovirus have been as high as 75% in a day care (Reintjes et al. 1999).
Coxsackie A21	• In a study of 39 volunteers, 10 individuals were directly inoculated with coxsackievirus A21 via an inhalant. Of these 10, 100% were infected and 80% showed signs of illness. Among the remaining 29 people who were deliberately exposed to the 10 inoculated volunteers as a result of living in the same enclosed room for 26 days, 100% became infected and 41% showed symptoms (Couch et al. 1969).
Coxsackie B2	• In a study of NPEV infection at a summer camp attended by institutionalized children and children with families, no mass spread of coxsackievirus B2 infection was observed even though children from families had only a 13% pre-summer antibody level for the virus, and one institutionalized child was known to have shed virus during the summer (Daneas et al. 1983).

Upon entry to the NYVW program, 40% of participants had antibodies for coxsackievirus A9 and B3, and 33% had antibodies for coxsackievirus B4. Seroprevalence rates for coxsackievirus B1–2 and B5 were 15%, 16%, and 15% respectively. It was also noted that the prevalence of coxsackievirus B2, B4, and B5 was low in the under-5 age group (Kogon et al. 1969). Immunity data for specific serotypes are summarized in Table D-2.

Shedding. Enteroviruses can be excreted in stools for over a month, but shedding from pharynx/upper alimentary tract generally lasts 1 to 2 weeks after infection (Melnick 1996). During the NYVW program, shedding of longer than one month was more common for coxsackievirus than echovirus, although shedding of echovirus lasted up to 42 days in one case. Coxsackievirus was shed up to 70 days. Nonetheless, the majority of shedding lasted for just one day in 83.8% of echovirus cases and 56.3% coxsackievirus cases (Kogon et al. 1969).

Table D-2. Immunity Data for Specific Enterovirus Serotypes

Echo11	A review by Jackson reports at least 40% of the adult U.S. population possesses neutralizing antibody to echovirus 11 (Jackson 1973).
Echo12	At the beginning of a study of echovirus 12 in volunteers, 417 (88%) lacked preexisting antibodies to the virus (Schiff et al. 1984).
Echo22	In 1995, a 1-month Finnish study that collected 110 serological samples from people aged 1–64 years found antibodies to echovirus 22 in 95% of neonates (presumably of maternal origin) but only 20% of infants between 2 and 10 months. Seroprevalence for echovirus 22 rises rapidly such that 91% of people over 1 year old had detectable antibody (Joki-Korpela and Hyypia 1998).
Echo 30	• In the same Finnish study as above, the seroprevalence of echovirus 30 was 38% in neonates, 10% in infants between 2 and 10 months, and 24% among those older than 1 year (Joki-Korpela and Hyypia 1998). • Prior to an epidemic of acute meningitis in Seattle, antibodies to echovirus 30 were detected in 7.2% of people participating in the Seattle Virus Watch Program (Hall et al. 1970).
Cox A21	A review of coxsackievirus A21 pointed to a study that showed that 18% of military recruits and 18% of civilians between ages 10–19 possessed antibody. Another study of 440 Californian volunteers found neutralizing antibody in 25%–30% of those tested (Jackson 1973).

According to Cooney, a causal relationship between infection and illness can be established for a cohort study if the onset of illness was 2 or 7 days after or 14 days before the first isolation. During the Seattle Virus Watch program, either echovirus or coxsackievirus was isolated from both respiratory and fecal samples up to 21 days before and after the onset of illness (Cooney et al. 1972). During the NYVW study, 43% of first enterovirus excretions occurred within –14 and +7 days of illness onset (Kogon et al. 1969). Among 372 children with enteroviral-like symptoms who were enrolled in a 4-month study at a private pediatric practice and were followed for 14 days, infections typically resulted in 7–10 day illnesses (Pichichero et al. 1998).

Person-to-Person Transmission. It has been reported that intrafamilial spread of NPEV is high and is related to the duration of shedding (Melnick 1996). In fact, in his review of enteroviruses, Melnick reports that the rate of infection among nonimmune family members or closed populations may reach as high as 80% (Melnick 1996).

Among the 93 members of the 19 SVW families that acquired infection during a 1968 outbreak of aseptic meningitis in Washington, 73% of members became infected (88% of children and 66% of adults) (Hall et al. 1970). The 64 non-SVW people with aseptic meningitis belonged to a total of 36 families. Secondary infection occurred in 48.5% of families within 2 weeks of the index case (Torphy et al. 1970).

Dose Response. Dose response is species- and serotype-specific for enteroviruses. Dose response investigations have been conducted for three of the enteroviruses; the findings are summarized in Table D-3.

Incubation. The incubation rate for enteroviruses ranges from 2-20 days but is usually 5-14 days and can last from 2 to 35 days (Melnick 1996). Specific incubation data are provided in Table D-4.

Table D-3. Dose-Response Studies for Specific Enterovirus Serotypes

Coxsackie A21	"Administration to 9 volunteers of enteric-coated capsules containing 10^5 coxsackievirus A21 particles produced no evidence of illness or serological response, although virus was recovered from 6" (Jackson 1973).
Echo11	"Eight volunteers given a 10^5-$TCID_{50}$ dose of echovirus 11 had no definite illness." The human ID_{50} was 10,000 monkey kidney $TCID_{50}$, and the extrapolated minimal infectious dose was less than 1/10,000 monkey kidney $TCID_{50}$ (Jackson 1973).
Echo12	A dose-response study of echovirus 12 in 149 healthy adult volunteers yielded an HID_{50} of 919 pfu (95% fiducial limits of 573–1503 PFU) and an HID_{01} of 17 PFU (95% fiducial limits of 1.0–56 PFU) (Schiff et al. 1984).

Table D-4. Incubation Data for Specific Enterovirus Serotypes

Echo 11	• 1-3 days (Moore 1982, Moore and Morens 1984) • Among 13 adult females inoculated with echovirus 11, the incubation period was 1-3 days. Illness subsided 6–7 days after the onset of symptoms (Jackson 1973)
Echo 20	The mean incubation period for echovirus 20 is 2.5 days (Moore 1982; Moore and Morens 1984).
Echo 30	During a community-wide outbreak of echovirus 30 associated with a swimming pool in which someone vomited, 94% of the 34 primary cases became ill within 72 hours of visiting the pool. The incubation rate of the 12 secondary household cases had a mean and median of 78 hours from exposure to the primary case in the same household (Kee 1994).
Cox A9	2-12 days (Moore 1982; Moore and Morens 1984)
Cox A21	In a 26-day U.S. study, 10 volunteers were deliberately inoculated with coxsackie A21 aerosol, and antibody was detected in all by day 3. All 29 non-inoculated individuals who were living in the same enclosed room as the ten inoculated volunteers also became infected beginning as early as day 4 and as late as day 24. The study found a mean incubation period to onset of shedding of 2.5 days (Couch et al. 1970).
Cox A24	12-30 hours, which is the shortest incubation period aside from enterovirus 70. (Moore 1982; Moore and Morens 1984)
Entero70	12-30 hours (Moore 1982; Moore and Morens 1984)

APPENDIX E

SOURCES FOR PARAMETER VALUES

	Parameter Description	Sources	Comments
α	Transition rate out of exposed state	1	Incubation periods are typically 7–14 d but can last up to 35 d. Some pathogens have shorter incubation periods (i.e., 12–30 h).
p_{sym}	Proportion of newly infected that will be symptomatic.	2	Generally, about 50% of infections are asymptomatic.
δ	Transition rate out of diseased state.	3	Varies with strain. Illness typically last 7–10 d in children. Other data suggest 1-21 d range. 1 to 2 wks in adults (data from one strain). Aseptic meningitis cases last about 3 d +/- 1 d.
σ_1,σ_2	Transition rate out of carrier state	4	Virus can be excreted from stool for over a mo. One study reported that half of those infected shed for 1 d. Shedding can persist for up to 70 d. Varies with strain.
γ	Transition rate out of postinfection state	5	Conflicting reports. Evidence that some protection exists. Protection is likely only partial, and antibodies do not remain in high concentrations for longer than 1 or 2 yr.
β_{s2}	Transmission rate due to person-to-person route.		Range set to explore model properties
ϕ	Shedding rate (organisms shed per person per day).		Virions per infected person per day.
k_w	Rate of movement of organisms from contamination site to exposure site.		Range set to explore model properties
μ	virus die off rate (natural mortality)		
β_{s1}	Transmission rate due to environment-to-person route.		β_{s1} is a function of the following 9 parameters
f	% pathogens from W_4 present in biosolids		Range set to explore model properties
i_b	Ingestion rate of treated soil	6	Case study assumes pica children not included.
i_e	Exposure through environmental routes	7	Used data that estimated ingestion rate from water
W_{ext}	External virus from environment		Because the model only simulates internal sources of pathogens this parameter = 0

[1] (Buckland 1961, Melnick 1996, Moore and Morens 1984)
[2] (Aoki and Sawada 1992, Buckland 1961, Couch et al. 1970, Hall et al. 1970, Melnick 1996)
[3] (Melnick 1996, Pichichero et al. 1998, Reintjes et al. 1999, Rice et al. 1995)
[4] (Couch et al. 1970, Hall et al. 1970, Kogon et al. 1969, Melnick 1996)
[5] (Aoki and Sawada 1992, Bern et al. 1992, Hall et al. 1970, Kogon et al. 1969, Schiff et al. 1984)
[6] (Calabrese et al. 1997b, Clausing et al. 1987, Hawley 1985, Lepow et al. 1975, Lepow et al. 1974, Roels et al. 1980, Stanek and Calabrese 2000)
[7] (USEPA 1989a)

Parameter		Description	Sources	Comments
V_b		Concentration of virus in sludge, from external source. Values range from 1 – 220 pfu/g	[8]	Values shown represent virus levels in secondary sludge. Since we are only simulating internal sources of pathogens this parameter is set to 0.
A		Dose-response parameter	[9]	Midrange of the 95% confidence interval for $\alpha = (0.126, 0.5)$
B		Dose-response parameter	[10]	Midrange of the 95% confidence interval for $\beta = (0.21, 0.84)$
X_e		Environmental attenuation factor		Range set to explore model properties
X_b		Biosolid attenuation factor (computed as the product of the 6 parameters listed below)		The low estimate is the product of all of the middle values, since $X_b=0$ is uninteresting.
	r	Treatment reduction. Log removal due to treatment (Depends on type of Tx and Tx factors; e.g. duration, temperature, dewaterating)	[11]	Case study conditions: Anaerobic digestion for approximately 20 days at 35° C, dewatered to 20% solids
	s	Storage reduction rate. Depends on Temperature, humidity, storage location (e.g. soil type/lagoon), soil condition, moisture content, UV radiation, strain of microorganism.	[12]	Case study: Air dried, northern CA climate during dry season. Ahmed and Sorensen data based on seeded biosolids much higher than Reimers data from lagoon storage
	t_s	Time of storage prior to land application		90 d is assumed in case study
	a	Application reduction. Depends on type of application (surface, tilled into soil) and environmental conditions (wind, rain, etc.)		When land applied to surface, a=1. Varies depending on application method (e.g. tilling into soil).
	l	Die off rate (depends on soil type, humidity, temp, UV exposure, moisture content)	[13]	Case Study: northern CA climate during dry season. General agreement among cited references
	t_l	Time to exposure	[14]	503B Rule specifies 1 year restricted access after Class B application

[8] (Albert and Schwartzbrod 1991; Goddard et al. 1981; Mignotte et al. 1999; Pederson 1981; Soares et al. 1994; Sorber and Moore 1986; Tata et al. 2000a)
[9] (Teunis et al. 1996)
[10] (Teunis et al. 1996)
[11] (Ahmed and Sorensen 1997; Feachem 1983; Sorber and Moore 1986; Tata et al. 2000b)
[12] (Ahmed and Sorensen 1995; Reimers et al. 1989)
[13] (Bitton et al. 1984; Farrah et al. 1981; Farrah and Bitton 1981; Hurst et al. 1979; Sorber and Moore 1986; Tierney et al. 1977)
[14] (USEPA 1995)

APPENDIX F

CART Tree from Simulations Discussed in Chapter 3.0

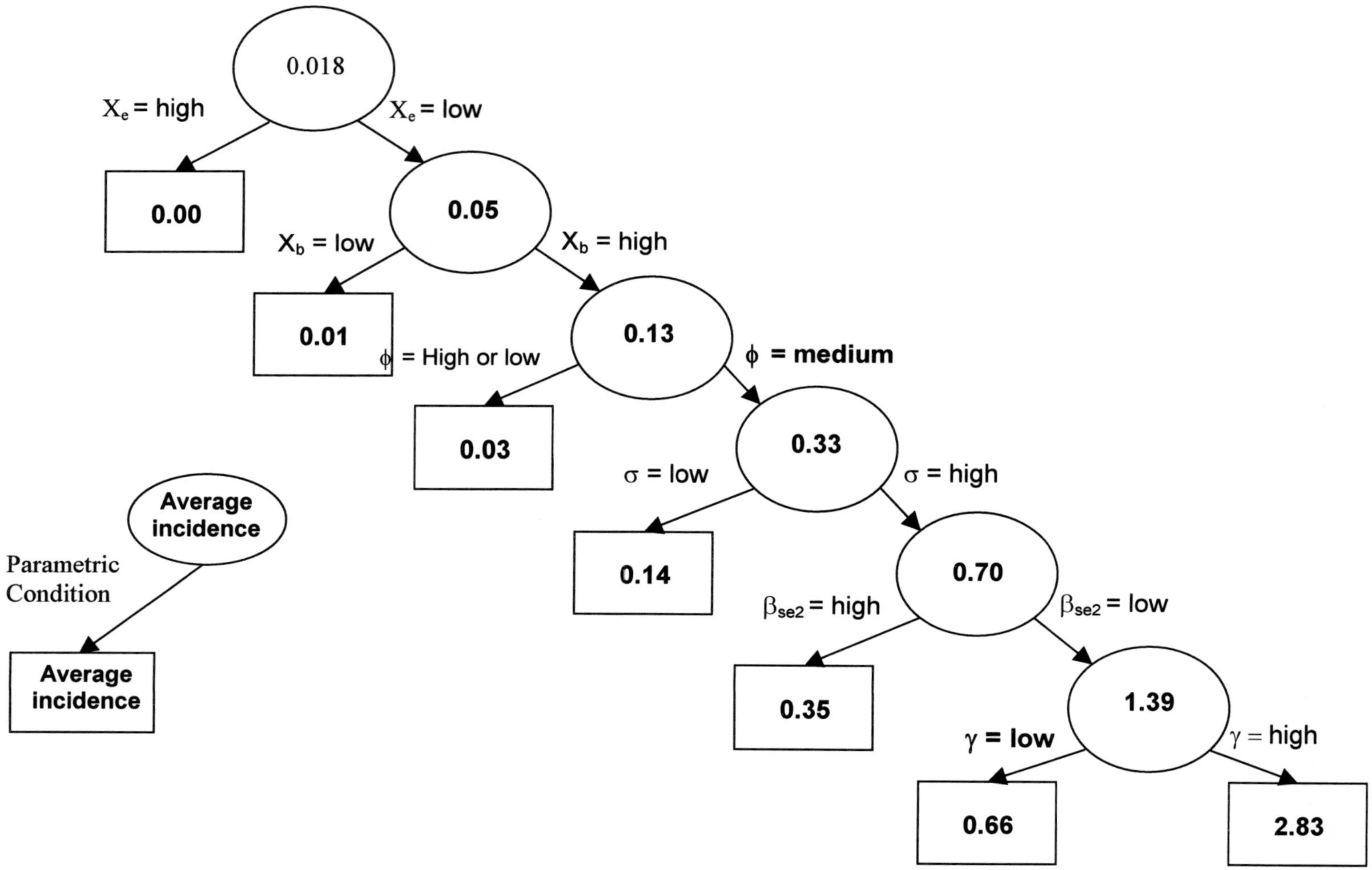
0.018
Xe = high
Xe = low
0.00
0.05
Xb = low
Xb = high
0.01
0.13
ϕ = High or low
ϕ = medium
0.03
0.33
σ = low
σ = high
0.14
0.70
βse2 = high
βse2 = low
0.35
1.39
γ = low
γ = high
0.66
2.83
Average incidence
Parametric Condition
Average incidence

REFERENCES

Ahmed, A. U., and D. L. Sorensen. 1995. Kinetics of Pathogen Destruction During Storage of Dewatered Biosolids. *Water Environ Res* 67: 143-150.

—. 1997. Autoheating and pathogen destruction during storage of dewatered biosolids with minimal mixing. *Water Environ Res* 69: 81-94.

Akin, E. W., and J. C. Hoff 1978. *Human viruses in the aquatic environment: A status report with emphasis on the EPA Research Program.* U.S. Environmental Protection Agency, Cincinnati, Ohio.

Albert, M., and L. Schwartzbrod. 1991. Recovery of Enterovirus from Primary Sludge Using Three Elution Concentration Procedures. *Water Sci Technol* 24: 225-228.

Anderson, M. A., M. H. Stewart, M. V. Yates, and C. P. Gerba. 1998. Modeling the impact of body contact recreation on pathogen concentrations in a source water drinking water reservior. *Water Res* 32: 3293-3306.

Aoki, K., and H. Sawada. 1992. Long-term observation of neutralization antibody after enterovirus 70 infection. *Jpn J Ophthalmol* 36: 465-8.

Bern, C., M. A. Pallansch, H. E. Gary, Jr., J. P. Alexander, T. J. Torok, R. I. Glass, and L. J. Anderson. 1992. Acute hemorrhagic conjunctivitis due to enterovirus 70 in American Samoa: Serum-neutralizing antibodies and sex-specific protection. *Am J Epidemiol* 136: 1502-6.

Binder, S., D. Sokal, and D. Maughan. 1986. Estimating soil ingestions. The use of tracer elements in estimating the amount of soil ingested by young children. *Arch Environ Health* 41: 341-345.

Bitton, G., O. C. Pancorbo, and S. R. Farrah. 1984. Virus transport and survival after land application of sewage sludge. *Applied Environ Microbiol* 47: 905-909.

Breiman, L., J. Freidman, R. Olshen, and C. Stone. 1984. *Classification and regression trees.* Wadsworth Intl. Group, Pacific Grove, Ca.

Brookhart, M. A., A. H. Hubbard, M. J. Vanderlaan, J. M. Colford, and J. N. S. Eisenberg. In press. Statistical estimation of parameters in a disease transmission model: Analysis of a *Cryptosporidium* outbreak. *Statistics in Medicine.*

Buckland, F. E. 1961. Inoculation of human volunteers with echovirus type 20. *British Medical Journal*: 397-400.

Calabrese, E. J., and E. J. Stanek. 1991. A guide to interpreting soil ingestion studies II. Qualitative and quantitative evidence of soil ingestion. *Regul Toxicol Pharmacol* 13: 263-277.

Calabrese, E. J., E. J. Stanek, R. C. James, and S. M. Roberts. 1997a. Soil ingestion: A concern for acute toxicity in children. *Environ Health Perspect* 105: 1354-1358.

Calabrese, E. J., E. J. Stanek, P. Pekow, and R. M. Barnes. 1997b. Soil ingestion estimates for children residing on a Superfund site. *Ecotoxicity Environ Safety* 36: 258-268.

CDC. 1997. Nonpolio enterovirus surveillance — United States, 1993-1996. *MMWR* 46: 748-750.

Chick, S. E., J. S. Koopman, S. Soorapanth, and M. E. Brown. 2001. Infection transmission system models for microbial risk assessment. *Sci Total Environ* 274: 197-207.

Clausing, P., B. Brunekreef, and J. H. van Wijnen. 1987. A method for estimating soil ingestion by children. *Int Arch Occup Environ Health* 59: 73-82.

Cooney, M. K., C. E. Hall, and J. P. Fox. 1972. The Seattle virus watch. 3. Evaluation of isolation methods and summary of infections detected by virus isolations. *Am J Epidemiol* 96: 286-305.

Couch, R. B., R. G. Douglas, Jr., K. M. Lindgren, P. J. Gerone, and V. Knight. 1970. Airborne transmission of respiratory infection with coxsackievirus A type 21. *Am J Epidemiol* 91: 78-86.

Couch, R. B., V. Knight, P. J. Gerone, T. R. Cate, and R. G. Douglas. 1969. Factors influencing response of volunteers to inoculation with Coxsackie virus A type 21. *Am Rev Resp Dis* 99: 24-30.

Crabtree, K. D., C. P. Gerba, J. B. Rose, and C. N. Haas. 1997. Waterborne adenovirus: A risk assessment. *Water Sci Technol* 35: 1-6.

Daneas, L., M. Havelková, and I. Procházková. 1983. Prevalence of infection with nonpolio enteroviruses among healthy children from a children's home and in a summer camp. *J Hygiene Epidemiol Microbiol Immunol* 27: 163-72.

Eisenberg, J. N., E. Y. W. Seto, A. W. Olivieri, and R. C. Spear. 1996. Quantifying water pathogen risk in an epidemiological framework. *Risk Analysis* 16: 549-563.

Eisenberg, J. N. S., E. Y. W. Seto, J. Colford, A. W. Olivieri, and R. C. Spear. 1998. An analysis of the Milwaukee *Cryptosporidium* outbreak based on a dynamic model of disease transmission. *Epidemiology* 9: 255-263.

EOA. 1995. *Microbial risk assessment for reclaimed water*. Prepared for the Irvine Water District and the National Water Resource Institute.

—. 1999. *Evaluation of pathogen risk assessment framework.* Prepared by J. A. Soller, J. N. Eisenberg, and A. W. Olivieri for ILSI Risk Science Institute.

Farber, J. M., W. H. Ross, and J. Harwig. 1996. Health risk assessment of *Listeria monocytogenes* in Canada. *Int J Food Microbiol* 30: 145-156.

Farrah, S., G. Bitton, E. Hoffmann, O. Lanni, O. C. Pancorbo, M. Lutrick, and J. Bertrand. 1981. Survival of enteroviruses and coliform bacteria in a sludge lagoon. *Applied Environ Microbiol* 41: 459.

Farrah, S. R., and G. Bitton. 1981. Fate of viruses following land application of sewage sludge. II. In M. Goddard and M. Butler, eds. *Viruses and wastewater treatment*. Pergamon Press, Oxford, U.K.

Faust, J. G., and R. L. Oberst. 1996. Proving its worth: Economic value of biosolids to farmers. *Biocycle* 37: 67-69.

Feachem, R. G. 1983. *Sanitation and disease: Health aspects of excreta and wastewater management.* John Wiley & Sons, Chichester, U.K.

Gibbs, R. A., C. J. Hu, G. E. Ho, P. A. Phillips, and I. Unkovich. 1995. Pathogen die-off in stored wastewater sludge. *Water Sci Technol* 31: 91-95.

Goddard, M. R., J. Bates, and M. Butler. 1981. Recovery of indigenous enteroviruses from raw and digested sewage sludges. *Appl Environ Microbiol* 47: 1023-1028.

Haas, C. N. 1983. Estimation of risk due to low doses of microorganisms: A comparison of alternative methodologies. *Am J Epidemiol* 55: 573-82.

Haas, C. N., J. B. Rose, and C. P. Gerba. 1999. *Quantitative microbial risk assessment*. J. W. Wiley, Inc., New York.

Hall, C. E., M. K. Cooney, and J. P. Fox. 1970. The Seattle virus watch program. I. Infection and illness experience of virus watch families during a communitywide epidemic of echovirus type 30 aseptic meningitis. *Am J Pub Health Nations Health* 60: 1456-65.

Hawley, J. K. 1985. Assessment of health risk from exposure to contaminated soil. *Risk Analysis* 5: 289-302.

Hay, J. C. 1996. Pathogen destruction and biosolids composting. *Biocycle* 37: 67+.

Hennekens, C. H., J. E. Buring, and S. L. Mayrent. 1987. *Epidemiology in medicine*. Little Brown, Boston.

Hu, C. J., R. A. Gibbs, and G. E. Ho. 1997. Evaluation of culture media for detection of salmonellae in composted biosolids. *Water Res* 31: 2664-2667.

Hurst, C., W. Benton, and R. Stetler. 1989. Detecting viruses in water. *J Am Water Works Assoc* 81: 71-80.

Hurst, C. J., S. Farrah, C. Gerba, and J. Melnick. 1979. Development of quantitative methods for the detection of enteroviruses in sewage sludges during activation and following land disposal. *Applied Environ Microbiol* 37: 626.

ILSI. 1996. A conceptual framework to assess the risks of human disease following exposure to pathogens. *Risk Analysis* 16: 841-848.

Jackson, G. G. 1973. Viruses causing common respiratory infections in man. Enteroviruses and paramyxoviruses. *J Infect Dis* 128: 387-408.

Joki-Korpela, P., and T. Hyypia. 1998. Diagnosis and epidemiology of echovirus 22 infections. *Clin Infect Dis* 26: 129-36.

Kee, F. 1994. A community outbreak of echovirus infection associated with an outdoor swimming pool. *J Pub Health Med* 16: 145-8.

Kogon, A., I. Spigland, T. E. Frothingham, L. Elveback, C. Williams, C. E. Hall, and J. P. Fox. 1969. The virus watch program: A continuing surveillance of viral infections in metropolitan New York families. VII. Observations on viral excretion, seroimmunity, intrafamilial spread and illness association in coxsackie and echovirus infections. *Am J Epidemiol* 89: 51-61.

Koopman, J. S., and I. M. Longini. 1994. The ecological effects of individual exposures and nonlinear disease dynamics in populations. *Am J Pub Health* 84: 836-842.

Koopman, J. S., I. M. Longini, J. A. Jacquez, C. P. Simon, D. G. Ostrow, W. R. Martin, and D. M. Woodcock. 1991. Assessing risk factors for transmission of infection. *Am J Epidemiol* 133: 1199-1209.

LaGoy, P. K. 1987. Estimated soil ingestion rates for use in risk assessment. *Risk Analysis* 7: 355-359.

Lepow, M., L. Bruckman, M. Gillette, S. Markowitz, R. Rubino, and J. Kapish. 1975. Investigations into sources of lead in the environment of urban children. *Environ Res* 10: 415-426.

Lepow, M., L. Bruckman, R. Rubino, S. Markowitz, M. Gillette, and J. Kapish. 1974. Role of airborne lead in increased body burden of lead in Hartford children. *Environ Health Perspect* 7: 99-102.

Logan, T. J. 1995. Gaining public acceptance for beneficial use of biosolids. *Biocycle* 36: 61-64.

McDonald, A. C., W. R. Mac Kenzie, D. G. Addiss, M. S. Gradus, G. Linke, E. Zembrowski, M. R. Hurd, M. J. Arrowood, P. J. Lammie, and J. W. Priest. 2001. *Cryptosporidium parvum*-specific antibody responses among children residing in Milwaukee during the 1993 waterborne outbreak. *J Infect Dis* 183: 1373-9.

Melnick, J. L. 1996. Enteroviruses: Polioviruses, coxsackieviruses, echoviruses, and newer enteroviruses. In B. N. Fields, Knipe, D. M., Howley, P. M., et al., ed. *Fields virology,* third edition. Lippincott-Raven Publishers, Philadelphia.

Mignotte, B., A. Maul, and L. Schwartzbrod. 1999. Comparative study of techniques used to recover viruses from residual urban sludge. *J Virol Methods* 78: 71-80.

Monpoeho, S., A. Dehee, B. Mignotte, and J. Schwartzbrod. 2000. Quantification of enterovirus RNA in sludge samples using single tube real-time RT-PCR. *Biotechniques* Jul, 29: 88-93.

Moore, M. 1982. Enteroviral diseases in the United States, 1970-1979. *J Infect Dis* 146: 103-108.

Moore, M., and D. Morens. 1984. Enteroviruses, including polioviruses. In R. Belshe, ed. *Textbook of human virology*. PSG Publishing Company, Littleton, Mass.

NRC. 1983. *Risk assessment in the federal government: Managing the process*. National Research Council, National Academy Press, Washington, D.C.

Pallansch, M. A. 1997. Coxsackievirus B epidemiology and public health concerns. *Curr Topics Microbiol Immun* 223: 13-30.

Pederson, D. 1981. Density levels of pathogenic microorganisms in municipal wastewater sludge: Indicator organisms in Sonoran Desert soil amended with sewage sludge. *Environ Sci Health* A28: 1287-1302.

Pichichero, M., S. McLinn, H. Rotbart, M. Menegus, M. Cascino, and B. Reidenberg. 1998. Clinical and economic impact of enterovirus illness in private pediatric practice. *Pediatrics* 102: 1126-34.

Reddy, K. R., R. Khaleel, and M. R. Overcash. 1981. Behavior and transport of microbial pathogens and indicator organisms in soils treated with organic wastes. *J Environ Qual* 10: 255-266.

Regli, S., J. B. Rose, C. N. Haas, and C. P. Gerba. 1991. Modeling the risk from *Giardia* and viruses in drinking water. *J Am Water Works Assoc* 83: 76-84.

Reimers, R. S., M. D. Little, T. G. Akers, W. D. Henriques, R. C. Badeaux, D. B. McDonnell, and K. K. Mbela. 1989. *Persistence of pathogens in lagoon stored sludge*. U.S. Environmental Protection Agency, Washington, DC. EPA/600/2-89/015, PB-89-190359.

Reintjes, R., M. Pohle, U. Vieth, O. Lyytikainen, H. Timm, E. Schreier, and L. Petersen. 1999. Community-wide outbreak of enteroviral illness caused by echovirus 30: A cross-sectional survey and a case-control study. *Pediatr Infect Dis J* 18: 104-8.

Rice, S., R. Heinl, L. Thornton, and S. Opal. 1995. Clinical characteristics, management strategies, and cost implications of a statewide outbreak of enterovirus meningitis. *Clin Infect Dis* 20: 931-7.

Riggle, D. 1996. Biosolids benefit agricultural land. *Biocycle* 37: 76-79.

Roels, H. A., J. P. Buchet, and R. R. Lauwerys. 1980. Exposure to lead by the oral and the pulmonary routes of children living in the vicinity of a primary lead smelter. *Environ Res:* 81-94.

Sanaa, M., N. Bemrah, S. Meyer, O. Cerf, and H. Mohammed. 2000. Quantitative risk assessment related to microbial food contamination. *Revue d'Epidemiologie et de Sante Publique* 48: 11-23.

Sartwell, P. 1950. The distribution of incubation periods of infectious disease. *Am J Hyg* 51: 310-318.

—. 1952. The incubation period of poliomyelitis. *Am J Pub Health* 42: 1403-1408.

—. 1966. The incubation period and the dynamics of infectious disease. *Amer J Epidemiol* 83: 204-216.

Schiff, G. M., G. M. Stefanovic, E. C. Young, D. S. Sander, J. K. Pennekamp, and R. L. Ward. 1984. Studies of echovirus-12 in volunteers: Determination of minimal infectious dose and the effect of previous infection on infectious dose. *J Infect Dis* 150: 858-866.

Simon, S. L. 1998. Soil Ingestion by Humans: a Review of History, Data, and Etiology with Application to risk assessment of radioactively contaminated soil. *Health Phys* 74: 647-672.

Soares, A., T. Straub, I. Pepper, and C. Gerba. 1994. Effect of anaerobic digestion on the occurrence of enteroviruses and *Giardia* cysts in sewage sludge. *J Environ Sci Health* A29: 1887-1897.

Sorber, C. A., and B. Moore. 1986. *Survival and transport of pathogens in sludge-amended soil: A critical literature review*. Office of Research and Development, U.S. Environmental Protection Agency, Washington, D.C. EPA/600/2-87/028. PB-87-18033.

Stanek, E. J., 3d, and E. J. Calabrese. 2000. Daily soil ingestion estimates for children at a Superfund site. *Risk Analysis* 20: 627-35.

Stanek, E. J., and E. J. Calabrese. 1995. Daily Estimates of Soil Ingestion in Children. *Environ Health Perspect* 103: 276-285.

Tata, P., C. Lue-Hing, J. Bertucci, S. Sedita, and G. Knafi. 2000a. Class A biosolids production by a low-cost conventional technology. *Water Environ Res* 72(4) 413-422.

Tata, P., C. Lue-Hing, and G. Knafi. 2000b. Statistical evaluation of pathogen inactivation for a conventional low-cost technology class A biosolids process. *Water Environ Res* 72(4) 423-431.

Tenenbaum, D. 1997. The beauty of biosolids. *Environ Health Perspect* 105: 32-6.

Teunis, P. F. M., O. G. van der Heijden, J. W. B. van der Giessen, and A. H. Havelaar. 1996. *The dose-response relation in human volunteers for gastrointestinal pathogens*. National Institute of Public Health and the Environment, Bilthoven, The Netherlands. 284550002.

Tierney, J. T., R. Sullivan, and E. P. Larkin. 1977. Persistence of poliovirus I in soil and on vegetables grown in soil previously flooded with inoculated sewage sludge or effluent. *Appl Environ Microbiol* 33: 109-113.

Torphy, D. E., C. G. Ray, R. S. Thompson, and J. P. Fox. 1970. An epidemic of aseptic meningitis due to echovirus type 30: Epidemiologic features and clinical and laboratory findings. *Am J Pub Health Nations Health* 60: 1447-55.

U.S. Environmental Protection Agency. 1989a. *Exposure factors handbook,*. U.S. Environmental Protection Agency, Washington, D.C. EPA/600/8-89/043.

—. 1989b. *Interim final guidance on soil ingestion rates*. U.S. Environmental Protection Agency, Washington, D.C. OWPE/OERR, DN: 9850.4.

—. 1989c. *Pathogen risk assessment for land application of municipal sludge. Volume I: Methodology and computer model*. U.S. Environmental Protection Agency, Washington, D.C. PB90-17190.

—. 1991a. *Preliminary risk assessment for bacteria in municipal sewage sludge applied to land.* Office of Health and Environmental Assessment, U.S. Environmental Protection Agency, Washington, D.C. EPA/600/S6-91/006.

—. 1991b. *Preliminary risk assessment for parasites in municipal sewage sludge applied to land.* Office of Health and Environmental Assessment, U.S. Environmental Protection Agency, Washington, D.C. EPA/600/S6-91/001.

—. 1992. *Preliminary risk assessment for viruses in municipal sewage sludge applied to land.* Office of Health and Environmental Assessment, United States Environmental Protection Agency, Washington DC. EPA/600/R-92/064.

—. 1993a. *A plain English guide to the EPA part 503 biosolids rule*. U.S. Environmental Protection Agency, Washington, D.C. EPA/832/R-93/003.

—. 1993b. *Standards for the use or disposal of sewage sludge, final rules*. U.S. Environmental Protection Agency, Washington, D.C. 40 CFR Part 257.

—. 1994. *Land application of sewage sludge*. U.S. Environmental Protection Agency, Washington, D.C. EPA/831-B-93-002b.

—. 1995. *A guide to the biosolids risk assessment for the EPA part 503 rule*. U.S. Environmental Protection Agency, Washington, D.C. EPA/832-B-93-005.

Van Wijnen, J. H., P. Clausing, and B. Brunekereef. 1990. Estimated soil ingestion by children. *Environ Res* 51: 147-62.

Voysey, P. A., and M. Brown. 2000. Microbiological risk assessment: A new approach to food safety control. *Int J Food Microbiol* 58: 173-179.

2002 WERF Subscribers

The following agencies and corporations are WERF's charter stormwater research supporters:

Stormwater Agency

California
Monterey, City of
Santa Rosa, City of

Colorado
Boulder, City of

Georgia
Griffin, City of

Kansas
Overland Park, City of

Kentucky
Louisville & Jefferson County Metropolitan Sewer District

North Carolina
Cary, Town of
Charlotte, City of/Mecklenburg County

Pennsylvania
Philadelphia, City of

Corporate

BaySaver Inc.
Vortechnics Inc.
WWETCO, LLC

State

Fresno Metropolitan Flood Control District, Calif.
Urban Drainage & Flood Control District, Colo.

Wastewater Utility

Alabama
Montgomery Water Works & Sanitary Sewer Board

Alaska
Anchorage Water & Wastewater Utility

Arizona
Gila Resources
Glendale, City of, Utilities Department
Mesa, City of
Phoenix Water Services Department
Pima County Wastewater Management

Arkansas
Little Rock Wastewater Utility

California
Central Contra Costa Sanitary District
Contra Costa Water District
Crestline Sanitation District
Delta Diablo Sanitation District
Dublin San Ramon Services District
East Bay Municipal Utility District
El Dorado Irrigation District
Escondido, City of
Fairfield-Suisun Sewer District
Irvine Ranch Water District
Las Virgenes Municipal Water District
Lodi, City of
Los Angeles, City of
Los Angeles County, Sanitation Districts of
Orange County Sanitation District
Palo Alto, City of
Riverside, City of
Sacramento Regional County Sanitation District
San Diego Metropolitan Wastewater Department, City of
San Francisco, City & County of
San Jose, City of
Santa Barbara, City of
Santa Rosa, City of
South Bayside System Authority
South Orange County Wastewater Authority
Union Sanitary District

Colorado
Boulder, City of
Colorado Springs, City of
Littleton/Englewood Water Pollution Control Plant
Metro Wastewater Reclamation District, Denver

Connecticut
New Haven, City of, WPCA

District of Columbia
District of Columbia Water & Sewer Authority

Florida
Broward, County of
Fort Lauderdale, City of
Gainesville Regional Utilities
JEA
Kissimmee, City of, Department of Water Resources
Miami-Dade Water & Sewer Authority
Orange County Utilities Department
Orlando, City of
Reedy Creek Improvement District
Seminole County Environmental Services
St. Petersburg, City of
Stuart Public Utilities
Tallahassee, City of
Tampa, City of
West Palm Beach, City of

Georgia
Atlanta Wastewater Services
Augusta, City of
Clayton County Water Authority
Cobb County Water System
Columbus Water Works
Fulton County
Gwinnett County Department of Public Utilities
Macon Water Authority
Savannah, City of

Hawaii
Honolulu, City and County of

Illinois
American Bottoms Wastewater Treatment Plant
Dupage, County of
Greater Peoria Sanitary District
Metropolitan Water Reclamation District of Greater Chicago
Wheaton Sanitary District

Indiana
Fort Wayne, City of
Indianapolis, City of

Iowa
Ames, City of
Des Moines Metro Wastewater Reclamation Authority

Kansas
Johnson County Unified Wastewater Districts
Unified Government of Wyandotte County/Kansas City, City of

Kentucky
Louisville & Jefferson County Metropolitan Sewer District

Louisiana
Sewerage & Water Board of New Orleans

Maine
Bangor, City of

Maryland
Anne Arundel County Bureau of Utility Operations
Howard County Department of Public Works
Washington Suburban Sanitary Commission

Massachusetts
Boston Water & Sewer Commission
Massachusetts Water Resources Authority
Upper Blackstone Water Pollution Abatement District

Michigan
Ann Arbor, City of
Detroit, City of

Grand Rapids, City of
Holland Board of Public Works
Lansing, City of
Owosso Mid-Shiawassee County WWTP
Saginaw, City of
Wayne County Department of Environment
Wyoming, City of

Missouri
Independence, City of
Kansas City Missouri Water Services Department
Little Blue Valley Sewer District
Metropolitan St. Louis Sewer District

Nebraska
Lincoln Wastewater System

Nevada
Henderson, City of

New Jersey
Passaic Valley Sewerage Commissioners

New York
New York City Department of Environmental Protection

North Carolina
Cary, Town of
Charlotte/Mecklenburg Utility Department
Durham, City of
Metropolitan Sewerage District of Buncombe County
Orange Water & Sewer Authority

Ohio
Akron, City of
Butler County Department of Environmental Services
Columbus, City of
Metropolitan Sewer District of Greater Cincinnati
Northeast Ohio Regional Sewer District

Oklahoma
Tulsa, City of

Oregon
Clean Water Services
Eugene/Springfield Water Pollution Control
Water Environment Services

Pennsylvania
Philadelphia, City of
University Area Joint Authority, State College

South Carolina
Charleston Commissioners of Public Works
Mount Pleasant Waterworks & Sewer Commission
Spartanburg Sanitary Sewer District

Tennessee
Cleveland, City of
Knoxville Utilities Board
Murfreesboro Water & Sewer Department
Nashville Metro Water Services

Texas
Amarillo, City of
Austin, City of
Dallas Water Utilities
Denton, City of
El Paso Water Utilities
Fort Worth, City of
Gulf Coast Waste Disposal Authority
Houston, City of
San Antonio Water System
Trinity River Authority

Utah
Salt Lake City Corporation

Virginia
Alexandria Sanitation Authority
Arlington, County of
Fairfax County Virginia
Hampton Roads Sanitation District
Henrico, County of
Hopewell Regional Wastewater Treatment Facility
Loudoun County Sanitation Authority
Prince William County Service Authority
Richmond, City of

Washington
Edmonds, City of
Everett, City of
King County Department of Natural Resources
Seattle Public Utilities

Wisconsin
Green Bay Metro Sewerage District
Madison Metropolitan Sewerage District
Milwaukee Metropolitan Sewerage District
Racine, City of
Sheboygan Regional Wastewater Treatment
Wausau Water Works

Australia
Sydney Water Corporation

Canada
Toronto, City of, Ontario

Mexico
Servicios de Agua y Drenaje de Monterrey, I.P.D.

United Kingdom
Yorkshire Water Services Limited

Corporate

The ADVENT Group Inc.
Alan Plummer & Associates
Alpine Technology Inc.
Aquateam–Norwegian Water Technology Centre A/S
Black & Veatch
Boyle Engineering Corporation
BPR CSO
Brown & Caldwell
Burns & McDonnell
The Cadmus Group
Camp Dresser & McKee Inc.
Carollo Engineers Inc.
Carpenter Environmental Associates Inc.
CDS Technologies Inc.
Chemtrac Systems Inc.
CH2M HILL
Clancy Environmental Consultants Inc.
Damon S. Williams Associates, LLC
Earth Tech Inc.
Ecolab Water Care Services
EMA Inc.
The Eshelman Company Inc.
Finkbeiner, Pettis, & Strout Inc.
Frontier Geosciences Inc.
ftn Associates Inc.
Gannett Fleming Inc.
GE Betz
Golder Associates Inc.
Greeley and Hansen LLC
The HACH Company
Hazen & Sawyer, P.C.
HDR Engineering Inc.
HNTB Corporation
HydroQual Inc.
Insituform Technologies Inc.
Institute for Environmental Technology & Industry, Korea
Jacobson Helgoth Consultants Inc.
Jason Consultants Inc.
Jordan, Jones, & Goulding Inc.
KCI Technologies Inc.
Kelly & Weaver, P.C.
Kennedy/Jenks Consultants
Komline Sanderson Engineering Corporation
Lawler, Matusky & Skelly Engineers, LLP
Limno-Tech Inc.
Lombardo Associates Inc.
Malcolm Pirnie Inc.
McKim & Creed
MWH
Odor & Corrosion Technology Consultants Inc. (OCTC)
ONDEO Degremont Inc.
PA Government Services Inc.
Parametrix Inc.
Parsons
Post, Buckley, Schuh & Jernigan
The RETEC Group
R.M. Towill Corporation
Royce Technologies
Sear-Brown
SYNAGRO
Tetra Tech Inc.
Trojan Technologies Inc.
URS Corporation
USFilter NATC
Wade-Trim Inc.
Weston Solutions Inc.
Woodard & Curran
Woodruff & Howe Environmental Engineering Inc. (WHEE)
WRc/D&B, LLC

Industry

American Electric Power
BP-Amoco Corporation
Chevron Research & Technology Company
The Coca-Cola Company
Dow Chemical Company
DuPont Company
Eastman Chemical Company
Eastman Kodak Company
Merck & Company Inc.
ONDEO Services
Procter & Gamble Company
Reliant Energy
Severn Trent Services Inc.
Shell Global Solutions
Thames Water Plc
United Water Services LLC

WERF Product Order Form

As a benefit of joining the Water Environment Research Foundation, subscribers are entitled to receive one complimentary copy of all final reports and other products. Additional copies are available at cost (usually $10). To order your complimentary copy of a report, please write "free" in the unit price column. WERF keeps track of all orders. If the charge differs from what is shown here, we will call to confirm the total before processing.

Name ______________________ Title ______________________

Organization ______________________

Address ______________________

City ______________ State ________ Zip Code ________ Country ________

Phone ______________ Fax ______________ Email ______________

Stock #	Product	Quantity	Unit Price	Total
			Postage & Handling	
			VA Residents Add 4.5% Sales Tax	
			Canadian Residents Add 7% GST	
			TOTAL	

Method of Payment: (All orders must be prepaid.)

❑ Check or Money Order Enclosed

❑ Visa ❑ Mastercard ❑ American Express

Account No. ______________________ Exp. Date ________

Signature ______________________

Shipping & Handling:

Amount of Order	United States	Canada & Mexico	All Others
Up to but not more than:	Add:	Add:	Add:
$20.00	$5.00*	$8.00	50% of amount
30.00	5.50	8.00	40% of amount
40.00	6.00	8.00	
50.00	6.50	14.00	
60.00	7.00	14.00	
80.00	8.00	14.00	
100.00	10.00	21.00	
150.00	12.50	28.00	
200.00	15.00	35.00	
More than $200.00	Add 20% of order	Add 20% of order	

*minimum amount for all orders

Note: Please make checks payable to the Water Environment Research Foundation.

To Order (Subscribers Only):

Log on to www.werf.org and click on the "Product Catalog."

Phone: (703) 684-2470
Fax: (703) 299-0742.

WERF
Attn: Subscriber Services
601 Wythe Street
Alexandria, VA 22314-1994

To Order (Non-Subscribers):

Non-subscribers may be able to order WERF publications either through WEF (www.wef.org) or IWA Publishing (www.iwapublishing.com). Visit WERF's website at www.werf.org for details.